THE PHYSICS OF MUSIC

Readings from

**SCIENTIFIC
AMERICAN**

THE PHYSICS
OF MUSIC

with an Introduction by
Carleen Maley Hutchins

 W. H. Freeman and Company
San Francisco

Note on cross-references to SCIENTIFIC AMERICAN *articles:* Articles included in this book are referred to by title and page number; articles not included in this book but available as Offprints are referred to by title and offprint number; articles not included in this book and not available as Offprints are referred to by title and date of publication.

Library of Congress Cataloging in Publication Data

Main entry under title:

The physics of music.

Bibliography: p.
Includes index.
1. Music—Acoustics and physics—Addresses, essays, lectures. I. Hutchins, Carleen Maley. II. Scientific American.
ML3805.P4   781'.22   77–28461
ISBN 0–7167–0096–4
ISBN 0–7167–0095–6 pbk.

Printed in the United States of America

9 8 7 6 5 4 3 2 1

CONTENTS

THE PHYSICS OF MUSIC

INTRODUCTION

Music has been with us for thousands of years. The human voice must have been the first instrument exploited for the making of music, and primitive man surely also used drumming, whistling, and rattling to make pleasing sounds. In many parts of the world, archeologists have found instruments made of such materials as stone, metal, and bone that are many thousands of years old. By the time of the rise of the early cultures in the Near East, the Babylonian and the Egyptian, the plucked string of the hunter's bow may have evolved into an early form of the harp to be used with pipes and drums of various sorts in ceremonials.

The first discovery in musical acoustics was made by Pythagoras in the sixth century B.C. Knowing nothing of frequencies of vibration, he noticed that the musical consonances we call the *octave, fifth,* and *fourth* occur when he "stopped" a given string of fixed tension respectively at one-half, two-thirds, and three-quarters of its length. For Pythagoras this was a fact of mystical numerology rather than of physics. The discovery that the corresponding vibration rates are inversely proportional to these fractions came more than 2000 years after Pythagoras' observation. Since early medieval times, the octave, fifth, and fourth have been called perfect consonances, and until the end of the Middle Ages, these intervals provided the theoretical basis for almost all Western polyphonic music. Though music has been with us from the oldest times, the development of the science of music had to await the discovery that there exists a relationship between frequency of vibration and tone.

Early in the history of Western music, various instruments, particularly the strings, were used to double or support parts in vocal music. Gradually the individuality of the instruments themselves began to be appreciated, and in the sixteenth century a specifically instrumental idiom began to emerge. This change in idiom and the change in the locations of performance from churches to halls of noble residences—and, eventually, to concert auditoriums and opera houses—led inevitably to a change in the instruments needed. History shows that instrument makers sought to meet the prevailing demands by improving the mechanisms of instruments, by refining their sounds, by extending their compass and dynamic range, and even by inventing completely new instruments. At that time, the best instrument makers were inspired experimentalists gifted with a great deal of intuition rather than trained theoreticians. This is particularly true of some of the violin makers of the late seventeenth and early eighteenth centuries such as Antonio Stradivari and Giuseppe Guarneri, who made instruments whose quality has scarcely been excelled.

Progress in musical acoustics was much slower than the development of music and instruments and concert halls. Some of the general principles of acoustics were discovered by such early scientists as Galileo Galilei (1564–1642), who described the phenomenon of sympathetic vibrations or resonance and the frequency of pendulum vibration based on the length of the pendulum; Marin Mersenne (1588–1648), who is credited with the first correct published account of the vibrations of strings and their frequencies; Robert Hooke (1635–1703), who connected frequency of vibration with pitch; Joseph Saveur (1653–1716), who laid the foundation for the concept of the fundamental and harmonic overtones, which was later developed into the celebrated theorem of Fourier (1768–1830), based on the principle of the coexistence of small oscillations known as the *superposition principle;* J. L. Lagrange (1736–1813), who not only solved the problems of the vibrating string in elegant analytical fashion, but is known to have worked on the sounds produced by organ pipes and musical wind instruments in general; Ernst F. F. Chladni (1756–1824), who described a method of using sand sprinkled on solid elastic vibrating plates to show the nodal lines, whose exact forms, however, defied analysis for many years; and S. D. Poisson (1781–1840), who at the same time worked on the analogous problem of the vibration of a flexible membrane such as the drum head. Although this is by no means a complete list of early researchers in acoustics, it is interesting to note that apparently the early acoustician did not pay very much attention to musical instruments. One exception was Félix Savart (1791–1841), who applied the principles of Chladni's vibrating plates to violins, constructing a trapezoidal violin for experimental purposes, and outlined many of the basic principles of violin acoustics. In his *Treatise on the Construction of Bowed String Instruments* (1819), Savart stated a challenge that is still valid for violin research today: "It is to be presumed that we have arrived at a time when the efforts of scientists and those of artists are going to unite to bring to perfection an art which for so long has been limited to blind routine."

It was not until the middle of the nineteenth century that Hermann von Helmholtz (1821–1894), Lord Rayleigh (John William Strutt, 1842–1919), and others provided the theoretical foundation for modern acoustics. Two milestones were Rayleigh's *Theory of Sound* (1878) and Helmholtz's *On the Sensations of Tone as a Physiological Basis for the Theory of Music* (1863).

Despite these important theoretical contributions, only minor experimental progress was made in musical acoustics during the nineteenth century because equipment for processing and analyzing acoustical signals was cumbersome and often less responsive to subtle variations than were the musical instruments themselves. Not until the early twentieth century was there a constant flow of work in musical acoustics. This flow was initially scant but grew steadily; it was largely stimulated by improved electronic, acoustical, and optical testing equipment sensitive enough to measure the physical subtleties of musical instruments and the sounds they produce. During the last half century this stream of activity has enlarged enormously with increasingly sophisticated tape recorders, spectrum analyzers, lasers, computers, and other sophisticated devices. The availability of this hardware and the development of new methods of analysis have made possible a better understanding of all musical instruments. In the brass instruments and woodwinds, in which the vibrating element, air, has simple and dependable properties, outstanding improvements have been possible. The increased dynamic range and improved playing qualities of these instruments have greatly enhanced the sounds of our symphony orchestras and popular music groups. At the same time, the development of the bowed string instruments has lagged, not because the science was being neglected, but because the technology of flexural vibrations in wood—a nonstandardizable material, but the best material available—is further removed from its scientific basis than the technology of air columns is from its science.

Some recent developments in the violin family, however, can help to change this, but acceptance of new ideas is slow. One is a method of testing that tells the violin maker when to stop thinning the arched top and back plates of the instrument before assembly to ensure that the tone of the completed violin is of excellent quality[1]. Another is a graduated octet of new instruments of the violin family spanning the entire musical range. These instruments, which have been developed according to a consistent theory of acoustics, are described in detail in "The Physics of Violins." Another new development is a fabricated material tailored to have the same acoustical properties as fine spruce wood, which has been used traditionally for instrument soundboards for hundreds of years. This is a sandwich construction of two very thin layers of oriented graphite fibers in an epoxy matrix with a layer of a fibrous material in between[2]. Experimental guitars and a violin with excellent tone have been made with this material, instead of the traditional spruce, as the top soundboard.

Current research on the physics of music that involves the human communication chain of composer–player–instrument–listener is becoming a truly multidisciplinary effort. This requires creative thinking in disciplines as widely separated as musical composition and performance, psychoacoustics, linguistics, materials research, architectural acoustics, and vibration analysis. Today there is increasing communication between workers in these areas on an international scale through exchange of publications, meetings, personal contacts, and correspondence. Thus, specifically focused research efforts in laboratories in Sweden, Germany, United Kingdom, Australia, Canada, and the United States are being related to a consistent framework through the Acoustical Society of America and the Catgut Acoustical Society, so that substantial progress is being made in important musical problems.

These readings from *Scientific American* are concerned primarily with the orderly relation between the physical properties of musical instruments and the sounds coming from them. Even the final article, "Architectural Acoustics," can be considered in this way if the listening area or music room is thought of as an instrument of sound transmission. Progress in the physics of music, like progress in most scientific fields, comes from the curious interweaving of seemingly unrelated ideas illuminated by gleams of intuition and the creative use of chance discoveries. The stimulating interaction between the art and the science of music is, in human terms, a lively exchange among experimental physicists, musicians, and instrument makers. It has already led to many new ideas and practical developments, some of which are explained in this Reader. Many more are yet to come.

The first article, "Physics and Music," provides an introduction to the other seven. It treats with the concept of music, the sound and acoustics of the various instrumental families, as well as those of the singing voice, and it includes basic explanations of harmonics, complex vibrations, resonances, Chladni patterns, and the ear as a detector of sound. "The Acoustics of the Singing Voice" gives a physical description of why and how a trained singer can produce sounds that are heard distinctly in a large hall over a full-scale symphony orchestra. "The Physics of the Piano" describes the development of the piano and its present mechanism, as well as some of the work being done on analysis and synthesis of musical tones. "The Physics of Wood Winds"

1. Hutchins, C. M., 1977. Another piece of the plate tap tone puzzle. *Catgut Acoustical Society Newsletter*, No. 28, page 22.
2. Haines, D. W., and N. Chang, 1975. Application of graphite composition in musical instruments. *Catgut Acoustical Society Newsletter*, No. 23, page 13.

and "The Physics of Brasses" clearly identify the technical problems of (and differences between) these two instrumental families and show the way in which modern physical theory is being applied to understanding and improving wind instruments. "The Physics of Violins" introduces the structure and function of the violin as an amplifier and radiator of vibrations from the bowed string and describes a new family of eight "violins," from treble to contrabass, that have been designed and built according to a consistent theory of acoustics. This is followed by "The Physics of the Bowed String," a discussion of the elementary physics of this unique generator of sound and the complications involved in understanding its behavior.

Progress in musical acoustics has been greatly stimulated by the theoretical and technological advances of recent years, but there is much yet to be learned about the traditional musical instruments that are the product of human ingenuity and pleasurable sensory experience over the centuries. Today new challenges for the creation and understanding of musical sounds are opening up through the use of sophisticated modern technologies and new research tools combined with the creative imagination of fine musicians.

Physics and Music

by Frederick A. Saunders
July 1948

The agreeable sound of simple melodies and Beethoven symphonies is guided by physical rules, plus a little physiology and psychology. The understanding of these principles can enhance musical creation and enjoyment

ANYONE who looks upon a great bridge arching across a wide river is thrilled by its beauty, and aware at the same time that a great deal of measuring, testing and calculating must have gone into its planning to make the structure safe. A bridge is an obvious combination of art and science. Not so obvious is the physical architecture of great music. One who listens to a symphony at an orchestral concert may know that the composer drew on his inspiration to fill pages with symbols, and that the conductor and his musicians interpret these to help bring to life again what was in the composer's mind. The listener is intellectually and emotionally moved by the sequence of sounds coming to him from many different sorts of instruments. But what has this bewilderingly complex example of art to do with science?

The answer is simple enough. Music is based on harmony, and the laws of harmony rest on physics, together with a little psychology and physiology. The simplest and most pleasant intervals of music have always existed among the harmonics of pipes and strings. From them grew the study of harmony, and they have formed the basis of many noble melodies. A classic example is the opening melody of Beethoven's *Eroica* symphony, whose first part consists of the simplest possible intervals flowing one after the other. Such simple combinations do something to our ears which is fundamentally pleasant and satisfying. Some musical instruments were well developed long before the subject of musical acoustics was born. Today the physics of music helps to guide improvements in musical instruments, in the construction of buildings with good acoustics, in the reproduction of music for immense audiences, and in many other ways.

To examine the physical basis of music we begin by considering the nature of sound. Sound is a word used in at least two senses: (1) the sensation produced in the brain by messages from the ear, and (2) the physical events outside the ear.

SYMPHONY is a vast blend of frequencies from many instruments. At left: Leopold Stokowski conducts rehearsal of New York Philharmonic.

The context usually makes it plain which meaning is intended. Thus we avoid long arguments over whether a sound can exist if there is no one present to hear it. Sound has its origin in a vibrating body, and the vibration may be *simple* or *complex*. The motion of the pendulum of a clock represents a simple vibration, one which is not rapid enough to be audible. To be heard as a musical tone, a vibration must have a frequency of at least 25 cycles per second. A pure tone is represented by a smooth

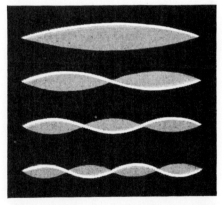

HARMONIC series is defined in various vibrations of a string. Harmonizing frequencies are two, three, four or more times simplest vibration (*top*).

curve in which distances to the right stand for time, and distances up and down correspond to the displacement of the vibrating body from its position of rest. A vibration of this sort is often called simple periodic motion because it repeats itself regularly with a constant period of time for each repetition. But pure musical tones are rare; the tones that are produced by musical instruments are almost always complex.

Complex vibrations can always be regarded as made up of a combination of simple vibrations of different frequencies. Their forms are very varied, as shown in the illustration on page 120. Sometime when you are out walking and have nothing better to do, try swinging your arms at different rates. The simplest case is easy: right arm going at twice the rate of the left. It is not quite so simple to make the right arm alone combine both of

these motions, and it is still harder to combine rates whose ratio is one to three, two to three, and so on. One gives up before long; yet any violin string can do this easily without becoming confused. It can combine as many as 20 different rates at the same time into one complex vibration, which is caused in this case by the complicated motion of the string under the bow. These frequencies are simply related; their values are proportional to the integers 1, 2, 3, 4 and so on. They form a harmonic series. The vibration with the lowest frequency. corresponding to the number 1, is called the fundamental; the sound with double this frequency is the first harmonic, and the higher harmonics are calculated in like manner.

I. Harmonic Analyzers

The scientific study of musical instruments depends partly upon the resolution of complex tones into their harmonic elements, a process called harmonic analysis. It is often of practical importance to determine what components are present in a tone and how strong each one is. One old method of analyzing a musical tone is to study its wave form, as pictured by means of a microphone, an amplifier, and a cathode-ray oscilloscope. But the wave is frequently very complicated, and its analysis by mathematical methods into the simple waves of which it is built is very slow and tedious. In recent years instruments have been developed which analyze complex tones automatically, yielding rapid and accurate results.

Some of these harmonic analyzers make use of the physical effect called resonance, which is a response produced in one body from the vibration of another body. It is easily demonstrated on a piano. In piano strings the harmonics are strong. If you press gently on the key an octave below middle C, so as to free the string but not to strike it, and then strike the middle C key sharply, you will hear a continuing middle C tone coming from the lower string. The experiment succeeds only if the strings are in tune. The middle C frequency (about 260 cycles per second) is

FREQUENCY RANGE of some musical instruments and other producers of sound is tabulated in chart adapted from book *The Psychology of Music*, by C. E. Seashore. Frequencies, noted in scale at the bottom of page, are plotted horizontally. Range of scale is 40 to 20,000 cycles, as compared with the human ear's approximate range of 25 to 30,000 cycles. The thin line within each light horizontal bar indicates actual range of frequencies produced by each method. Circles on each line indicate effective range estimated by a group of expert musicians. Vertical lines at the right end of each frequency line indicate range of associated noise. The instruments in black panel are, from top to bottom, tympani, snare drum, cello, piano, bass tuba, French horn, bassoon, clarinet, male speech, female speech and jingling keys. In blue panel are cymbals, violin, trumpet, flute and clapping hands.

equal to that of the first harmonic of the lower string; hence the lower one can respond.

By a variation of the experiment, one can play a chord on a single string. Hold the lower string open as before, but now give a strong impulse to three keys at once—middle C, the C above and the G between. After the upper strings have been quieted, all three tones will be heard coming from the lower string alone, which is resonating to three frequencies at once. This works as well the other way around: hold the same three upper keys open with the right hand and give the lower C a sharp blow. The three upper tones will be heard, coming from the three untouched strings. Or again, try singing a tone into a piano with the loud pedal pressed down. (This frees the strings to vibrate in resonance with any tone with which they agree in frequency.) When you stop singing, you will hear a faint mixture of tones issuing from the piano.

If we had some kind of attachment to the strings by means of which the response of each could be recorded, we should have one type of harmonic analyzer, but not a very good one. It would be unable to respond properly to frequencies lying between those of the strings. A more useful type of analyzer would be a single string whose pitch we could change slowly and steadily throughout the whole range of the musical scale. This could be fitted with an attachment which would record the string's responses, whenever they occurred, to the tone being analyzed. Such a device would be like the tuning apparatus in a radio receiver, which picks up radio waves on each frequency over which they are being broadcast. The device would miss nothing, but it would not be capable of making analyses instantaneously. The same sort of plan, carried out electrically, gives more rapid results. With suitable equipment it is possible to obtain within a few seconds a complete photographic analysis of a sustained tone, yielding numerical values for the strength and frequency of all harmonics present in the frequency range from 60 to 10,000 cycles per second. This method has been applied to the study of the tones of many instruments.

A remarkable frequency analyzer recently developed by R. K. Potter of the

Bell Telephone Laboratories gives a continuous analysis of speech; its result is appropriately called "visible speech." One speaks into a microphone and the oscillations of his speech are then passed through 12 electrical filters, each of which allows only a narrow range of frequency to pass. When amplified, each filtered set of oscillations lights a tiny "grain-of-wheat" lamp; there are 12 lamps, arranged vertically. The fundamental tone of the speech lights one lamp, the first harmonic another farther up, and so on. The lamps that light in response to the speaker indicate the frequencies present in his speech. To reproduce his speech pattern, the light from the lamps falls on a horizontal moving belt made of phosphorescent material, so arranged that each lighted lamp traces a separate luminous line on the belt. The result is a characteristic pattern for each vowel and consonant, defined by lines of varying frequency and duration. The accompanying illustration demonstrates how a phrase looks to the eye. A trained observer can read words and phrases at sight, and a person who has been deaf from birth may thus learn to read speech. He can also correct imperfections in his own speech by matching the patterns he produces against standard ones. This visible speech is exciting to watch, and it is likely to be of great help to the deaf.

II. The Violin

Now let us turn to the consideration of musical instruments, a subject in which harmonic analysis has been very useful. We may agree at the start that nothing deserves the name of musical instrument unless it can make a loud sound. Our greatest musical artists must fill large concert halls, and for this they need loud voices, violins, pianos or other instruments. Some musical instruments require a method of amplifying the vibrations created by the player to produce powerful tones. Consider the violin as an example.

A wire mounted on a bent iron rod, with no body or plate to shake, gives almost no sound when it is excited by bow or finger. The wire is too narrow to push the air about sufficiently to create a strong sound wave. Such a performance is analogous to trying to push a

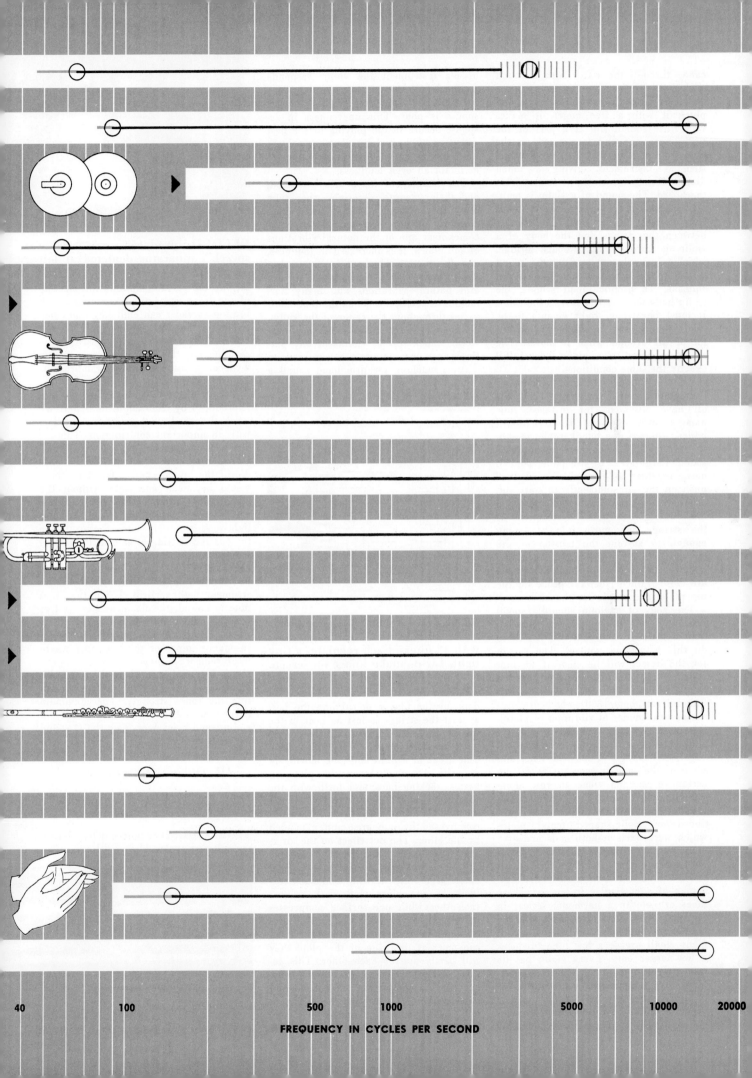

canoe through the water with a round stick as a paddle. If you stretch a piece of strong twine between two hands and pluck it with a free finger, it makes very little sound. But if a part of the twine near one end is pressed on the edge of a thin board, you have a crude stringed instrument, giving a much louder sound, which now comes from the board. The sound of a violin is emitted not from the strings or the bow but from its light wooden body. The contact between the strings and the body of a violin is through the wooden bridge, which is cleverly cut to filter the sound transmitted and remove some unpleasant squeaks. To produce loud sounds, the violin body must satisfy three conditions. It must be strong, light enough to be easily shaken, and big enough to push a lot of air around when it moves. The sounding board of a piano must fulfill exactly the same conditions.

As everyone knows, stiff objects vibrate much better than limp ones; we all have observed, for instance, how noisy a job it is to wrap a parcel in stiff paper, whereas if a handkerchief is substituted for the paper there is almost no sound. Large areas of stiff paper tend to move together as one piece, and thus push on the air sufficiently to start vigorous sound waves. In a violin, the wood must be light, so that the vibrations of the strings can move it, and strong enough to sustain the tension of the strings, which adds up to about 50 pounds. The kinds of wood most used are close-grained Norway spruce for the top plate, and maple for the back.

The body of a violin should respond equally to all frequencies of vibration within its range. The fact that it fails to do this is seldom noticed. The reason for the defect will be clear if we first consider the beautiful method devised by the German acoustical physicist Ernst Chladni (at about 1800) which discloses the natural modes of vibration of plates. By sprinkling sand on a flat metal plate and drawing a rosined bow across its edge, one can get a musical tone, and some of the sand is seen to move from certain areas and some to rest along quiet "nodal" lines. The accompanying illustrations show various figures produced on violin-shaped metal plates which were fixed at both ends and at a point corresponding to the violin sound post. In each figure there are several patterns, and each pattern is associated with a tone of a particular frequency. These tones are not in a harmonic series; in fact they are usually discordant with one another. A high tone forms a pattern of many small areas; a low tone produces a few larger ones. Every violin has its own natural modes of vibration, scattered over the musical scale, and eight or ten of them may be especially strong. When a violinist produces a tone coinciding with a strong natural frequency

of his instrument the violin responds loudly, but if he makes one in the range between two such frequencies, the response is poor. This unevenness in response occurs in the playing of the best artists on the best violins, but it is seldom noticed since no artist is expected to maintain an even loudness.

The number of harmonics produced, and their strength, determine the "tone color" or timbre of a sustained tone from a violin. Whenever one of the harmonics comes near one of the natural vibrations of the plates, it is increased in loudness, and the tone is changed in tone color. This happens often, because there are several natural vibrations and many harmonics in each tone. Thus the tone color varies throughout the range of the violin. No one tone color is characteristic of any violin—much less of violins of any particular age or from any one country.

As a machine for producing sound a

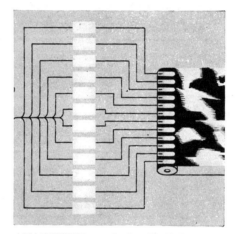

ANALYZER made by Bell Laboratories separates sound frequencies with 12 filters. Each regulates a tiny light. Lights make image on screen.

violin is very inefficient. Most of the work done by the player in rubbing the bow against the strings is lost as heat in the wood. The Chladni patterns show another reason for inefficiency. Two adjacent areas in a plate must be moving in opposite directions when the plate vibrates, rocking back and forth with the separating nodal line at rest between them. Thus at the same instant the air is compressed by one area and expanded by the other. The net effect on the air is greatly reduced, since the contributions of the two areas nearly cancel each other. Moreover, the front and back surfaces of any plate may work against each other: while one surface compresses the air, the back of the same area starts an opposite expansion. If the two waves can meet at the edge of the plate they will partly destroy each other. This action weakens the low tones particularly, not only in violins but in pianos and loud-speakers. To prevent this effect in loud-speakers, the vibrating area is commonly set into a "baffle" which, by en-

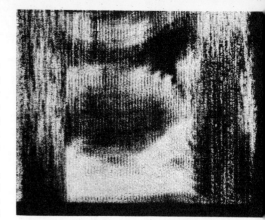

RECORD produced by "visible speech" apparatus depicted at left

larging the surface, inhibits the meeting of the front and back waves. Larger vibrating surfaces can emit low tones better. This is why the violoncello and double bass are made progressively bigger, and why the large sounding board in a concert grand piano helps to improve its deep bass tones.

Not all of the tone emitted by a violin is produced by vibration of its plates. We must also credit the air inside the box with an important contribution. This air can vibrate in and out of the f-holes with a frequency which lies in the middle of the lowest octave. The tone there would be mean and ugly without the added vibration of the inner air, as one can discover by plugging the f-holes lightly with cotton. When the air inside the box is vibrating at or near its natural frequency, its resonance is strong. This can be demonstrated by setting a candle in front of one of the holes, with the instrument held vertical. When the right note is bowed, the flame dances wildly; for all others it remains quiet. (The effect is most marked in a cello.) Air resonance improves the tone just where improvement is most needed; that is, over a few semitones where the small size of the violin prevents the body from emitting the tones strongly. The maximum effect is near C sharp on the G string in violins, and near A or B on the G string in violas and cellos.

III. Old v. New Instruments

Now what makes a superlative violin?

VIOLIN MUSIC recorded by visible speech apparatus shows a horizontal

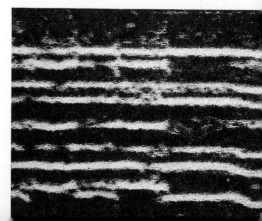

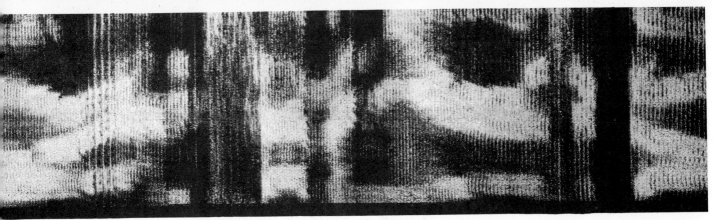

may be temporary image on a phosphorescent screen or, as in the illustration above, a pattern on a paper strip.

This pattern, which may be read by a trained observer, represents phrase "Four score and seven years ago . . ."

This question is endlessly debated, but it cannot be settled by arguments. The most accurate and careful measurements in a laboratory with modern equipment are required, and a start has already been made. The impression made by a violin on a listener is due to many features: the quality or "tone color" of sustained tones, the ease with which the tones begin, the rate of decay of the sound, the loudness in different parts of the range. These items are often lumped together under the word "tone"; here we must separate them carefully. The tone color of sustained tones is probably the least important of the lot. The loudness in different ranges of pitch may be the most vital consideration in the judgment of a violin. A bad violin is weak in the low tones and too strong in the squeaky top frequencies.

Old violins are almost always thought to be better than new ones, and European better than American. This opinion may come in part from psychological causes —our admiration of old civilizations, the influence of tradition and so on—but part of it certainly comes from the beauty of workmanship characteristic of the best old instruments, and from their rarity. It is as difficult for most violinists to find any defect in a Stradivarius as it is easy for them to criticize the best-made American violin. In recent years careful experiments have been made with excellent modern apparatus, seeking to measure all the mechanical features mentioned in the preceding paragraph. A great variation in values was found among 12 Strads, many other old

violins and a few dozen new ones, but the average values failed to show any consistent difference between old and new. This is not to say that there are no differences, but that the results were the same within the limits of error in the measurements, using very sensitive methods. These bold statements are supported by many "blindfold" audience tests, as well as by variations in professional opinions as to the merits of certain famous violins.

Violins seem to become lighter and better when played for a century or two. The effect of age on instruments which are not played appears to be small. Changes in the physical character of a violin can come about through vibration and also from contact with players. After a period of use a violin usually weighs more because it has absorbed water vapor from the air around the player. This makes the wood expand across the grain; when not in use it dries out again and contracts. These changes may alter both the physical and chemical properties of the wood. Some day it may be possible to attain the effects of years by a quick treatment of the wood; promising work along this line is now in progress.

There are methods of mapping out the natural vibrations of a violin by exciting it electrically and measuring its response at every frequency. This yields a curve, called the response curve, by means of which violins can be compared. The inequalities in response at various frequencies are remarkable, in both old and new instruments. All good violins should

in the future have a certified response curve furnished with them when they are offered for sale.

IV. The Piano

Some of the statements which I have made about the violin apply equally to the piano. The piano's sounding board acts like the violin body. While the violin has not changed in the last century, the piano has seen constant improvements in the sounding board, the strings, the hammers and the key action. So loud has the instrument become that the vibrations now shake the floor and are sometimes transmitted through the solid structure of a building to unexpected distances. In apartment houses peace may sometimes be preserved with the neighbors by placing rubber pads between the piano legs and the floor.

New problems arose with the invention of the piano's key and hammer mechanism. The hammer must be light but strong, in order to act quickly and give powerful blows to the strings. The pads must be soft to avoid the production of strong high harmonics that a hard hammer creates. (One can almost convert a piano into a harpsichord by using a teaspoon for a hammer.) When a player hits a key on the piano, the action gives the hammer a throw; at the moment when the hammer strikes a string it is not connected with the key, but is flying freely. It is as if the player were throwing soft balls at the strings from a distance. Once the hammer is on its free way, the player can do nothing more to it. His only control is

band for each harmonic produced by the instrument. Large number of bands illustrates complex nature of

musical sounds. Record shows passage from Glazounov's "Concerto in A Minor," played by Jascha Heifetz.

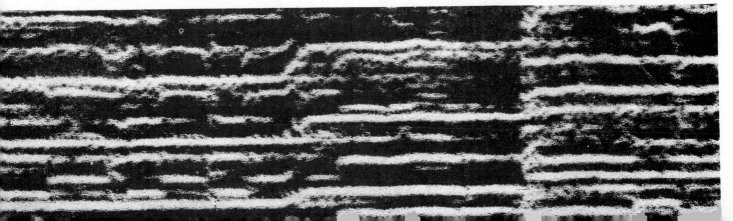

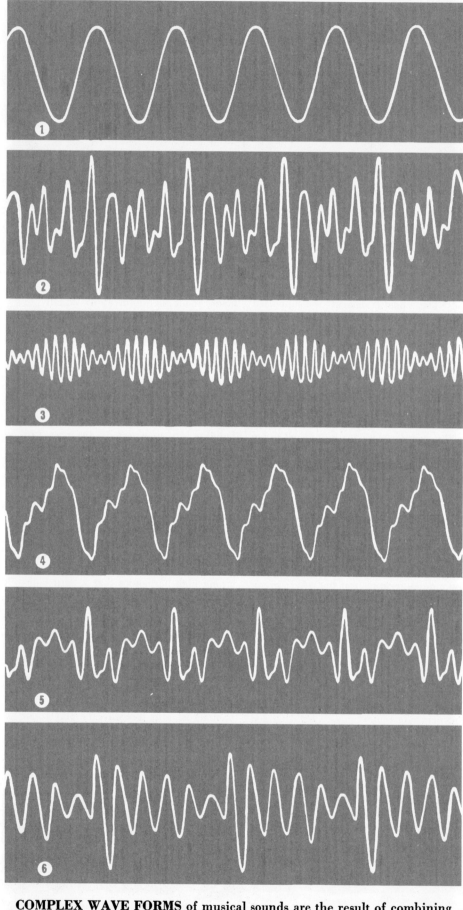

COMPLEX WAVE FORMS of musical sounds are the result of combining several simple forms. The forms above are (1) the simple tone of a tuning fork, (2) pure chord produced by four tuning forks struck together, (3) "beat" tone of two tuning forks with almost the same frequency. Characteristic instrumental forms were made by (4) violin, (5) oboe, (6) French horn.

through the initial speed he imparts to the hammer. Thus it is a fact that for a given hammer speed the tone is exactly the same whether the key is pressed by the finger of a great artist or by the tip of an umbrella. Any skeptic to whom this statement is repulsive should open up a piano and watch the motion of a hammer. Piano "touch" is of course a mixture of effects: besides hammer speed, which affects loudness and tone color, it depends on the sequence of tones, the length of time each key is held, the management of the pedals, the phrasing and so on. Of these the last three are perhaps the most important.

There are two subjects in musical acoustics, incidentally, which often arouse furious arguments. One is piano touch. The other is the alleged characteristic flavor of music in different keys. Pupils are often taught that D major is a martial key. Today military marches are played on a piano whose D is 294 cycles per second. A musician in Mozart's time would have had a D of about 278 cycles per second (our C sharp), since the pitch has risen about a semitone in this interval. If two performances of the same music in different pitches can produce the same impression, then the flavor of the key must come from its name and not from its pitch.

V. Wind Instruments

The wind instruments operate on a very different plan from the strings, and as sound-producers they are much more efficient. A stringed instrument loses considerable energy in transmitting its vibrations from the plate to the air; in a wind instrument the sound is emitted directly by vibrations of the air inside the pipe. Hence an instrument like the oboe or clarinet in the orchestra stands out against the string section, and two or three of them are considered sufficient to balance a much larger group of violins.

The sound waves in wind instruments are generated in a variety of ways: by thin streams of air issuing from slots (the organ) or from the player's lips (flute); by the vibrations of single reeds of cane (clarinet family), of double reeds of cane (oboe family), of metal (organ), or of the player's lips (cornet, horn). Except in the case of a metal reed, which has to be tuned to its pipe, the mechanism that excites a wind instrument has no very definite natural frequency but will accommodate itself to the rate of the vibration of the air in the pipe. This rate is determined by the time it takes an exciting air impulse, traveling with the speed of sound (about 1,100 feet a second), to go down the pipe and back. In instruments which have side holes, this wave is reflected not from the end of the pipe but from the first hole that is open. By this means the player controls the effective length of the pipe and the frequency or pitch of the sound produced. Shortening the pipe by opening successive holes makes it possible to produce the notes of the musical scale; the higher tones are obtained as harmonics of these fundamental vibrations. The sound of the flute comes from two holes, the one at the mouthpiece and the first open one lower down; the vibrating air dances in and out of these two holes simultaneously. The holes still lower down emit practically no sound. The same principles apply to the oboe or clarinet except that there is no hole in the mouthpiece. The lowest tone is the only one whose sound issues from the end of the instrument.

In the brass instruments, the length of the tube is governed either by a sliding piece (slide trombone) or by insertion of additional lengths of pipe by means of valves operated with the fingers. At each length a large series of harmonics can be blown, and with several lengths available all the notes of the scale can be played, many of them in more than one way. The fundamental tones are not often used.

The tone colors of wind instruments are not as variable as those of violins; hence the player's opportunities for virtuosity are more limited. On the organ, the only wind instrument that has separate pipes for each pitch, the organist can build up tone colors by combining pipes of the same pitch but different colors. In the brass instruments, a player produces a marked change in tone color when he puts his fist or some other object in the "bell" from which the sound comes. This muting of the tone corresponds to what happens when one loads the bridge of a violin with an extra weight.

VI. The Singing Voice

But none of these instruments has the variety of tone color available to a singer. The voice is the most versatile and expressive of all musical instruments.

The vocal cords vibrate somewhat as do the lips of a cornet player, that is, as a double reed. They produce a range of fundamental frequencies which is determined by the muscular tension that can be put on them and by their effective mass and length. The action of the cords has recently been photographed with a motion-picture camera, showing that they have a complicated, sinuous back-and-forth motion. Such a motion would be expected to generate a complex sound wave; voice sounds are indeed found to be rich in harmonics.

The throat and mouth space through which the sound passes on its way out can take the form of one chamber or be divided almost in two by the back of the tongue. A singer also varies the size of the mouth opening. These alterations enable the chamber to resonate to a variety of frequencies, some low as fundamentals, some high as harmonics. In singing, we presumably tune the chamber to resonate with the vocal cords at their fundamental

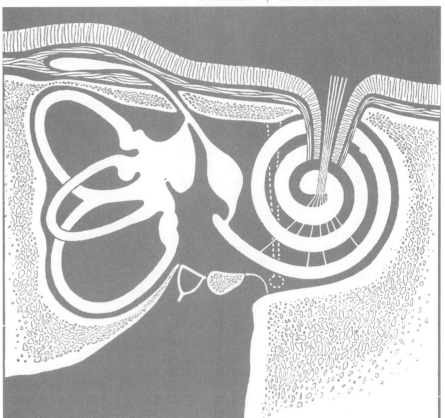

INNER EAR, here shown in highly diagrammatic drawing, is detector of sound. Spiral organ at right is the cochlea. From it run branches of the auditory nerve (*upper right*). These branches are attached to the basilar membrane, stretched across the cochlea's inside diameter along its full length. When sound vibrates membrane, nerve impulses are sent to the brain.

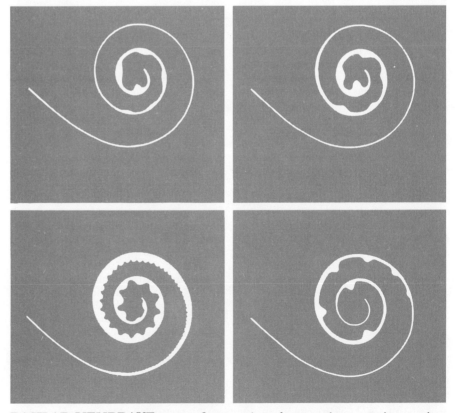

BASILAR MEMBRANE responds to various frequencies at various points along its length. Peaks on spiral diagrams show relative response. Two drawings at top show "false harmonics" of ear's response to a pure tone of increasing loudness. Two drawings below show membrane's accurate response to many harmonics of steamboat whistle (*left*) and note of a bugle (*right*).

or some harmonic. In general the pitch of the voice varies with the tension and length of the cords, its quality depends on the shape and size of the chamber, its loudness is determined by the amount of air pressure supplied by the lungs. The versatility of the voice comes from the ease and quickness with which all these changes can be made.

Singing teachers use certain special terms to describe all the processes involved in tone production. Although these terms have quite definite meanings to the teachers, to others such descriptions as "head tones," "chest register," and "tone placement" mean very little, and that little is probably misleading. One would suppose, for example, that the head and chest must vibrate somewhat at all times, and that the tone must always originate in the same place. One may also object to crediting the bony cavities in the head and the absorbent lung-space with helping to produce loud sounds, since these areas are powerless to contribute anything appreciable. It is to be hoped that before long there will be further experimental studies that will disclose the real behavior of the whole vocal apparatus, and that then such language can be used as will be understood by all.

VII. Musical Scales

There is one special study in which mathematics and music go hand in hand. This is in the construction of musical scales. People with unmusical ears sing up and down the range of pitches without hitting the same spot twice; but music cannot be built on this plan. The piano must have a pattern on its keyboard, and a fixed frequency for each key, as the flute has fixed positions for its side holes. The pattern of the keyboard repeats itself in each octave. An octave is measured by the first interval in the harmonic series. Two tones an octave apart have a frequency ratio of 2 to 1; they produce in our ears a simple motion and a pleasant impression. To produce a similarly pleasing effect within the octave, its intervals also should be simple, with ratios like the ones found in the harmonic series, such as the musical fifth (ratio 3 to 2) and the fourth (4 to 3). Thus the scale is built up on the plan of having as many pairs of tones as possible which please us when sounded together. At the same time the musician demands freedom to shift keys without running into any trouble with different sorts of intervals.

The final result is a scale of 12 notes with semitone intervals all exactly alike, and just filling an octave. The mathematician tells us that if we multiply the frequency of any starting note by the 12th root of two (or 1.05946), we obtain the frequency of the next higher note, and if we continue this process, after 12 multiplications we arrive at the beginning of the next octave. This scale does not give

us perfect musical intervals inside the octave, but there seems to be no way in which we can get a better one to fit all the conditions stated. The purist objects; he has a wonderful ear and he says it hurts to hear these intervals the least bit off. So the mathematician writes another paper on a perfect—but unusable—scale.

Recently the physicist and the psychologist have joined in the discussion. A new measuring device has been invented by O. L. Railsback, which he calls the chromatic stroboscope. With this he can measure the frequency of any tone while it is sounding, with a precision greater than we may ever need. It has been used to check the tuning of pianos. The results show that expert tuners agree among themselves but they tune the low notes too low and the high ones too high to fit the scale. They do this because it actually sounds better, and the explanation of this odd fact is that the harmonics of a piano string are themselves out of tune, and are

it does not always tell the strict truth. S. S. Stevens of Harvard University has shown that the pitch of a pure tone varies with its loudness. Low tones may drop a whole tone on the musical scale, while very high tones go the other way. If, while listening to a loud tone whose pitch is off, you cover your ears, the pitch goes back to where it belongs. Fortunately, since this effect is observed to an appreciable degree only for pure tones, it is of little importance in listening to most music, because the tones are complex. Moreover, at the pitch where the ear is most sensitive (2,000 cycles per second), the effect disappears.

The ear may even manufacture sounds that do not exist. Harvey Fletcher and his group at the Bell Telephone Laboratories have found that as the loudness of a pure tone increases, the ear begins to hear a change of tone color, seemingly caused by harmonics which appear in the tone in increasing number and strength. The tone increases in shrillness and harshness until

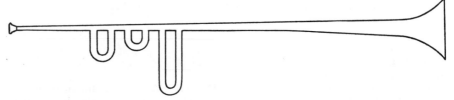

BRASS INSTRUMENT is stretched out to illustrate function of valves. Manipulating valves adds extra segments to effective length of pipe. This changes the rate of vibration of air in pipe and frequencies of its tones.

sharper than they should be. The 14th harmonic occurs about where the 15th belongs. The scale that results is no longer the exact scale of "equal temperament," which we have just considered, but a "spread" one whose octave ratio is slightly greater than two to one, while its fifths are almost true. All these years we have been using two scales without knowing it. To make matters worse, it has been shown that, in contrast to the piano, the harmonics of pipes and of bowed strings are not out of tune; so that the organ is presumably tuned in equal temperament. The violin is always tuned to perfect fifths, yet nobody minds when it is played with a piano tuned to a different scale. These strange differences seem to have escaped the notice even of our friend the purist.

Recently musical psychologists of the University of Iowa, under the leadership of Dr. C. E. Seashore, have measured the performance of a number of first-class professional singers and violinists, and found that they do not use the scale of equal temperament nor any other scale exactly. We must all be less sensitive to the refinements of tuning than was supposed. The scale (or scales) we now use is quite good enough for such ears as the best of us possess.

VIII. The Listening Ear

The ear, in fact, is a surprising organ;

it sounds like the blast of a cornet in one's ear. Yet an oscilloscope picture of the wave form of the sound shows no trace of these harmonics.

These ghostly harmonics arise somehow in the ear itself. The sensitive basilar membrane, where sound is detected by a series of nerve endings, has been proved to respond to different frequencies at different positions along its length. The membrane is spiral-shaped, and Fletcher pictures its "auditory patterns" by means of a set of spiral diagrams showing where disturbances occur in response to sounds of different pitch. In the case of a soft, pure tone, the membrane is disturbed only at the place appropriate to the frequency. But as the same tone grows louder, new disturbances mysteriously appear at the points where the harmonics of this tone would be recorded. The source of the false harmonics is probably traceable to a natural imperfection in the action of the mechanism of the middle ear.

A practical consequence of this quirk is that any tone, pure or complex, increases greatly in harshness as it becomes louder. Thus even a good radio gives a bad tone when turned up too loud; the ear is to be blamed, not the radio set. A violin has a harsher tone to the ear of the player than to a listener some distance away. A violin whose sound was amplified electrically to fill a large hall would sound quite unnatural.

IX. Room Acoustics

Science has made a very considerable contribution to music in connection with the acoustics of halls. To make clear the nature of this contribution we must consider some of the facts about sound in rooms. If the source of sound in a room is suddenly stopped, the sound lasts a little while; it dies down as it is absorbed or escapes through openings. The duration of this sound is long if the room is large or if the sound was a loud one; it is shortened if many absorbing substances are present. The absorptivity of a material is great if it is full of fine pores in which the regular vibrations that constitute sound are made irregular and thus turned into heat.

The best absorptive material is a closely packed audience. But porous plates of various sorts are available for covering walls or ceilings to cut down reflection of sound and increase absorption, in case the audience is not large enough. A bare room with hard walls reflects excellently, and this has two effects: the sound is made louder (just as white walls make a room lighter), and it is prolonged. Speech becomes hard to understand, because successive syllables overlap. Music usually benefits more by reflection than speech does: it has fewer short "syllables," and the reflections can make it loud enough to be heard well even in the rear seats of a very large hall.

Wallace Sabine of Harvard was the first to work out the proper way of correcting the acoustics of noisy halls by increasing their absorptivity. He founded architectural acoustics, which is fast becoming an exact science. It is now a simple matter to provide for good acoustics in a hall before it is built, and a bad hall can usually be made tolerable by treatment at any time.

One musical application of acoustics concerns the marked effect which the character of a room may have on the tone color and the loudness of a voice or other musical instrument. Most absorptive materials absorb more of the high tones than the low ones. When you select a piano in a bare showroom, it is likely to have a "brilliant" tone, meaning that it is strong in high frequencies. But if you place the same piano in a living room full of stuffed furniture, cushions and thick carpets, you may find its tone dull and weak. The high frequencies are still present, but they are quickly absorbed, and so you do not get the reinforcement of these tones that occurred in the showroom. A violin's tone color and power likewise depend on the sort of room in which it is played. On the other hand, a singer whose shrill high tones are hard to bear in an ordinary room should bring along a truckload of cushions, the presence of which would have the effect of greatly increasing the listeners' pleasure.

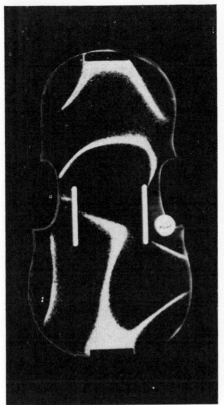
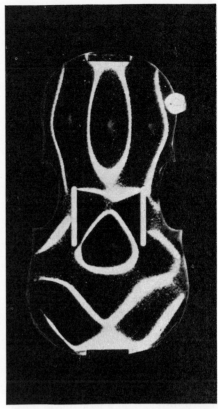

CHLADNI PLATES indicate the vibration of the body of a violin. These patterns were produced by covering a violin-shaped brass plate with sand and drawing a violin bow across its edge. When the bow caused the plate to vibrate, the sand concentrated along quiet nodes between the vibrating areas. Bowing the plate at various points, indicated by round white marker, produces different frequencies of vibration and different patterns. Low tones produce a pattern of a few large areas; high tones a pattern of many small areas. Violin bodies have a few such natural modes of vibration which tend to strengthen certain tones sounded by the strings. Poor violin bodies accentuate squeaky top notes. This sand-and-plate method of analysis was devised 150 years ago by the German acoustical physicist Ernst Chladni.

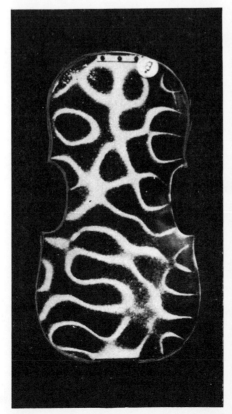
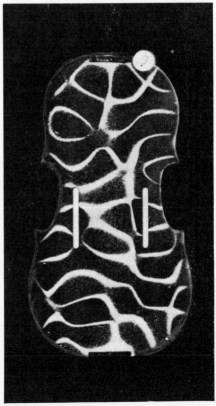

2

The Acoustics
of the Singing Voice

by Johan Sundberg
March 1977

The voice organ is an instrument consisting of a power supply (the lungs), an oscillator (the vocal folds) and a resonator (the larynx, pharynx and mouth). Singers adjust the resonator in special ways

Clearly there is something quite unusual about the voice of a first-class opera singer. Quite apart from the music, the intrinsic quality of such a voice can have a forceful impact on the listener. Moreover, a well-trained singer produces sounds that can be heard distinctly in a large opera house even over a high level of sound from the orchestra, and can do so week after week, year after year. If a second-rate singer or a completely untrained one tried to be heard over an orchestra, the result would be a scream and the singer's voice would soon fail. Is it only training that makes the difference? Or is the instrument that produces an excellent singer's voice itself different from other people's?

Let us begin with a description of that instrument. The voice organ includes the lungs, the larynx, the pharynx, the nose and the mouth. The main voice function of the lungs is to produce an excess of air pressure, thereby generating an airstream. The air passes through the glottis, a space at the base of the larynx between the two vocal folds (which are often called the vocal cords but are actually elastic infoldings of the mucous membrane lining the larynx). The front end of each vocal fold is attached to the thyroid cartilage, or Adam's apple. The back end of each is attached to one of the two small arytenoid cartilages, which are mobile, moving to separate the folds (for breathing), to bring them together and to stretch them. The vocal folds have a function apart from that of producing sound: they protect the lungs from any small objects entrained in the inspired airstream. Just above the vocal folds are the two "false" vocal folds, which are engaged when someone holds his breath with an overpressure of air in the lungs. The vocal folds are at the bottom of the tube-shaped larynx, which fits into the pharynx, the wider cavity that leads from the mouth to the esophagus. The roof of the

pharynx is the velum, or soft palate, which in turn is the door to the nasal cavity. When the velum is in its raised position (which is to say during the sounding of all vowels except the nasalized ones), the passage to the nose is closed and air moves out through the mouth.

The larynx, the pharynx and the mouth together constitute the vocal tract, a resonant chamber something like the tube of a horn or the body of a violin. The shape of the tract is determined by the positions of the articulators: the lips, the jaw, the tongue and the larynx. Movements of the lips, jaw and tongue constrict or dilate the vocal tract at certain sites; protruding the lips or lowering the larynx increases the length of the tract.

Now consider the voice organ as a generator of voiced sounds. Functionally the organ has three major units: a power supply (the lungs), an oscillator (the vocal folds) and a resonator (the vocal tract). With the glottis closed and an airstream issuing from the lungs, the excess pressure below the glottis forces the vocal folds apart; the air passing between the folds generates a Bernoulli force that, along with the mechanical properties of the folds, almost immediately closes the glottis. The pressure differential builds up again, forcing the vocal folds apart again. The cycle of opening and closing, in which the vocal folds act somewhat like the vibrating lips of a brass-instrument player, feeds a train of air pulses into the vocal tract. The frequency of the vibration is determined by the air pressure in the lungs and by the vocal folds' mechanical properties, which are regulated by a large number of laryngeal muscles. In general the higher the lung pressure is and the thinner and more stretched the vocal folds are, the higher is the frequency at which the folds vibrate and emit air pulses. The train of pulses produces a rapidly oscillating air pressure in the vocal tract;

in other words, a sound. Its pitch is a manifestation of the vibratory frequency. Most singers need to develop full control over a pitch range of two octaves or more, whereas for ordinary speech less than one octave suffices.

The sound generated by the airstream chopped by the vibrating vocal folds is called the voice source. It is in effect the raw material for speech or song. It is a complex tone composed of a fundamental frequency (determined by the vibratory frequency of the vocal folds) and a large number of higher harmonic partials, or overtones. The amplitude of the partials decreases uniformly with frequency at the rate of about 12 decibels per octave. The "source spectrum," or plot of amplitude against frequency, for a singer is not very different from that for a nonsinger, although the spectrum does tend to slope more steeply in soft speech than it does in soft singing.

The vocal tract is a resonator, and the transmission of sound through an acoustic resonator is highly dependent on frequency. Sounds of the resonance frequencies peculiar to each resonator are less attenuated than other sounds and are therefore radiated with a higher relative amplitude, or with a greater relative loudness, than other sounds; the larger the frequency distance between a sound and a resonance is, the more weakly the sound is radiated. The vocal tract has four or five important resonances called formants. The many voice-source partials fed into the vocal tract traverse it with varying success depending on their frequency; the closer a partial is to a formant frequency, the more its amplitude at the lip opening is increased. The presence of the formants disrupts the uniformly sloping envelope of the voice-source spectrum, imposing peaks at the formant frequencies. It is this perturbation of the voice-source envelope that produces distinguishable speech sounds: particular formant frequencies manifest themselves in the radiated spectrum as peaks in the envelope, and those peaks are characteristic of particular sounds.

The formant frequencies are determined by the shape of the vocal tract. If the vocal tract were a perfect cylinder closed at the glottis and open at the lips and 17.5 centimeters (about seven inches) long, which is about right for the average adult male, then the first four formants would be close to 500, 1,500, 2,500 and 3,500 hertz (cycles per second). Given a longer or shorter vocal tract, these basic frequencies are somewhat lower or higher. Each formant is associated with a standing wave, that is, with a static pattern of pressure oscillations whose amplitude is at a maximum at the glottal end and near a minimum at the lip opening [*see illustration on page 19*]. The lowest formant corresponds to a quarter of a wavelength, which is to say that a quarter of its wavelength fits within the vocal tract. Similarly, the second, third and fourth formants correspond respectively to three-quarters of a wavelength, one and a quarter wavelengths and one and three-quarters wavelengths.

Any change in the cross section of the vocal tract shifts the individual formant frequencies, the direction of the shift depending on just where the change in area falls along the standing wave. For example, constriction of the vocal tract at a place where the standing wave of a formant exhibits minimum-amplitude pressure oscillations generally causes the formant to drop in frequency; expansion of the tract at those same places raises the frequency.

The vocal tract is constricted and expanded in many rather complicated ways, and constricting it in one place affects the frequency of all formants in different ways. There are, however, three major tools for changing the shape of the tract in such a way that the frequency of a particular formant is shifted in a particular direction. These tools are the jaw, the body of the tongue and

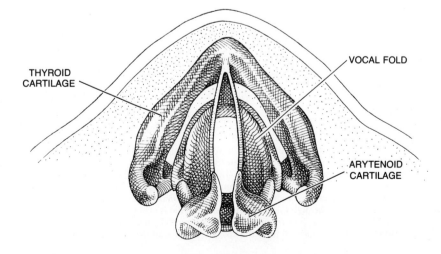

VOICE ORGAN is composed of the lungs and the larynx, pharynx, mouth and nose, shown in longitudinal section (*top*). The larynx is a short tube at the base of which are twin infoldings of mucous membrane, the vocal folds. The larynx opens into the pharynx; the opening is protected during swallowing by the epiglottis. The larynx, pharynx and mouth (and in nasal sounds also the nose) constitute the vocal tract. It is a resonator whose shape, which determines vowel sounds, is modified by changes in the position of the articulators: the lips, the jaw, the tip and body of the tongue and the larynx. The vocal folds, seen from above in a transverse section (*bottom*), are opened for breathing and are closed for phonation by the pivoting arytenoid cartilages.

Diagram labels (top): NASAL CAVITY, NASAL PHARYNX, SOFT PALATE, ORAL PHARYNX, EPIGLOTTIS, PHARYNX, LARYNX, FALSE VOCAL FOLD, LARYNGEAL VENTRICLE, VOCAL FOLD, ESOPHAGUS, LIPS, TONGUE, THYROID CARTILAGE, TRACHEA

Diagram labels (bottom): THYROID CARTILAGE, VOCAL FOLD, ARYTENOID CARTILAGE

the tip of the tongue. The jaw opening, which can constrict the tract toward the glottal end and expand it toward the lip end, is decisive in particular for the frequency of the first formant, which rises as the jaw is opened wider. The second-formant frequency is particularly sensitive to the shape of the body of the tongue, the third-formant frequency to the position of the tip of the tongue. Moving the various articulatory organs in different ways changes the frequencies of the two lowest formants over a considerable range, which in adult males averages approximately from 250 to 700 hertz for the first formant and from 700 to 2,500 hertz for the second. Moving the articulatory organs is what we do when we speak and sing; in effect we chew the standing waves of our formants to change their frequencies. Each articulatory configuration corresponds to a set of formant frequencies, which in turn is associated with a particular vowel sound. More specifically, the formant frequencies enhance voice-source partials of certain frequencies and thus manifest themselves as the peaks characterizing the spectrum envelope of each vowel sound.

All the elements and functions of the voice organ that I have been describing are common to singers and non-singers alike. Do singers bring still other faculties into play or manipulate the voice instrument in different ways? Let us begin by comparing normal male speech and operatic singing. Careful attention to a singer's voice reveals a number of modest but very characteristic deviations in vowel quality from those of ordinary speech. For example, the *ee* sound of a word such as "beat" is shifted toward the umlauted *ü* of the German "*für*"; the short *e* of "head" moves toward the vowel sound of "heard." The general impression is that the quality of the voice is "darker" in singing, somewhat as it is when a person yawns and speaks at the same time; voice teachers sometimes describe the effect as "covering."

These shifts in vowel quality have been found to be associated with peculiarities of articulation. In "covered" singing the larynx is lowered, and X-ray pictures reveal that the change in the position of the larynx is accompanied by an expansion of the lowest part of the pharynx and of the laryngeal ventricle, the space between the true vocal folds and the false ones. It is interesting to note that voice teachers tend to agree that the pharynx should be widened in singing, and some of them mention the sensation of yawning. In other words, a low larynx position and an expanded pharynx are considered desirable in singing.

What we recognize as a darkened voice quality in singing is reflected very clearly in the spectrum of a sung vowel sound. A comparison of the spectra of the vowel in "who'd" as it is spoken and sung shows that the two lowest formant frequencies are somewhat lower in the sung version and that the spectral energy, or amplitude, is considerably higher between 2,500 and 3,000 hertz [*see top illustration on page 21*]. This spectral-envelope peak is typical of all voiced sounds sung by professional male singers. Indeed, its presence, regardless of the pitch, the particular vowel and the dynamic level, has come to be consid-

OUTPUT SOUND

RADIATED SPECTRUM

AMPLITUDE →

FREQUENCY →

VOCAL TRACT (RESONATOR)

VOCAL-TRACT FREQUENCY CURVE

AMPLITUDE →

A B C

FREQUENCY →

VOICE SOURCE

VOICE-SOURCE SPECTRUM

AMPLITUDE →

FREQUENCY →

VIBRATING VOCAL FOLDS (OSCILLATOR)

AIRSTREAM

LUNGS (POWER SUPPLY)

VOICE ORGAN is composed functionally of a power supply, an oscillator and a resonator. The airstream from the lungs is periodically interrupted by the vibrating vocal folds. The resulting sound, the voice source, has a spectrum (*right*) containing a large number of harmonic partials, the amplitude of which decreases uniformly with frequency. The air column within the vocal tract has characteristic modes of vibration, or resonances, called formants (*A, B, C*). As the voice source moves through the vocal tract each partial is attenuated in proportion to its distance from formant nearest it in frequency. The formant frequencies thus appear as peaks in the spectrum of the sound radiated from the lips; the peaks establish particular vowel sounds.

ered a criterion of quality; the extra peak has been designated the "singing formant."

What is the origin of the singing-formant peak? The peaks in the spectrum envelope of a vowel normally stem, as I have explained, from the presence of specific formants. The insertion of an extra formant between the normal third and fourth formants would produce the kind of peak that is seen in the spectrum of a sung vowel [*see bottom illustration on page 21*]. Moreover, the acoustics of the vocal tract when the larynx is lowered are compatible with the generation of just such an extra formant. It can be calculated that if the area of the outlet of the larynx into the pharynx is less than a sixth of the area of the cross section of the pharynx, then the larynx is acoustically mismatched with the rest of the vocal tract; it has a resonance frequency of its own, largely independent of the remainder of the tract. The one-sixth condition is likely to be met when the larynx is lowered, because the lowering tends to expand the bottom part of the pharynx. I have estimated on the basis of X-ray pictures of a lowered larynx that this lowered-larynx resonance frequency should be between 2,500 and 3,000 hertz, that is, between the frequencies of the normal third and fourth

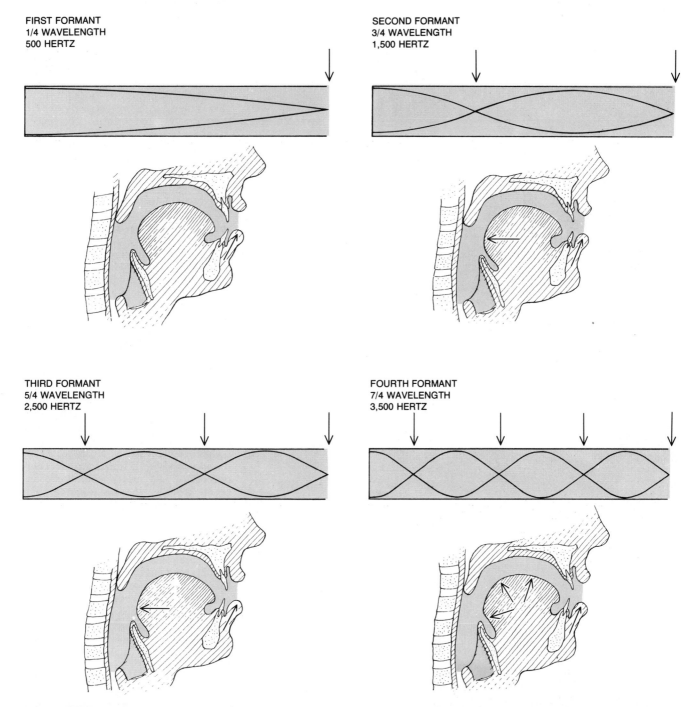

FIRST FORMANT
1/4 WAVELENGTH
500 HERTZ

SECOND FORMANT
3/4 WAVELENGTH
1,500 HERTZ

THIRD FORMANT
5/4 WAVELENGTH
2,500 HERTZ

FOURTH FORMANT
7/4 WAVELENGTH
3,500 HERTZ

FORMANTS correspond to standing waves, or static patterns of air-pressure oscillations, in the vocal tract. Here the first four formants are shown as standing waves in cylindrical tubes, the schematic equivalent of the vocal tract (*colored areas in drawings*). The sine waves represent the amplitude of the pressure differential, which is always maximal at the glottal end and minimal at the lips. For the lowest formant a quarter of a wavelength is within the vocal tract and, if the tract is 17.5 centimeters long, the formant's frequency is about 500 hertz (cycles per second). The second, third and fourth formants are 3/4, 5/4 and 7/4 of a wavelength, and their frequencies vary accordingly. If the area of the vocal tract is decreased or increased at a place where the formant's pressure amplitude is at a minimum (*arrows*), that formant's frequency is respectively lowered or raised; the same change in area has the opposite effect if it is at a pressure maximum.

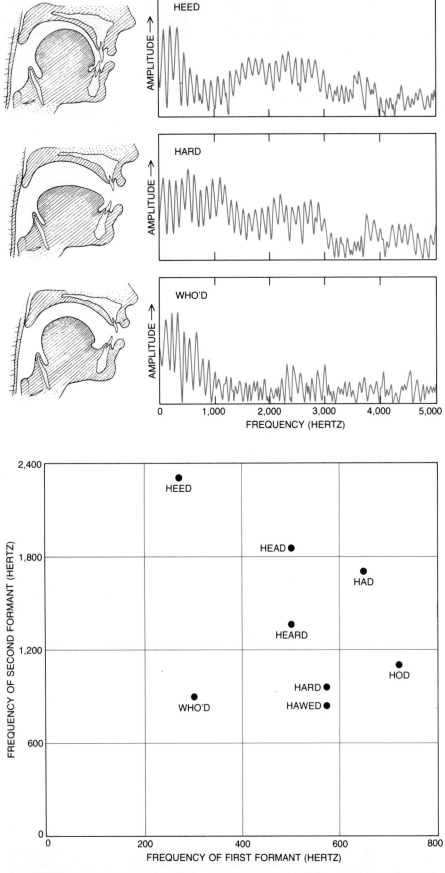

MOVEMENT OF ARTICULATORS changes the cross section of the vocal tract, shifting formant frequencies. Three articulatory configurations are shown (*top*) together with the spectrum of the vowel sound produced by each; the peaks in the spectrum envelope reflect the formant frequencies. The chart (*bottom*) gives the frequencies of the first and second formants in some English vowel sounds as spoken by an average male. For a female or a child the envelope pattern would be about the same but the peaks would be shifted somewhat higher in frequency.

formants and just where the singing-formant peak appears. The lowering of the larynx, in other words, seems to explain the singing-formant peak.

It also accounts for something else. Acoustically the expansion of the lowermost part of the pharynx is equivalent to an increase in the length of the vocal tract, and the lowering of the larynx adds still more to the length. The result is to shift downward all formant frequencies other than the larynx-dependent extra formant. This lowering of frequency is particularly notable in formants that depend primarily on the length of the pharynx. Two examples of such formants are the second formant of the vowels in "beat" and "head," and a drop in the frequency of those formants moves their vowels respectively toward those of "*für*" and "heard." The lowering of the larynx, then, explains not only the singing-formant peak but also major differences in the quality of vowels in speech and in singing.

To explain the singing formant's articulatory and acoustic origin is not enough, however. Why, one wonders, is it desirable for singers to lower the larynx, producing the singing formant and darkening the quality of their vowels? A plausible answer to the question has been found. It is related to the acoustic environment in which opera and concert singers have to work: in competition with an orchestra. Analysis of the average distribution of energy in the sounds of an opera or symphony orchestra shows that the highest level of sound is in the vicinity of 450 hertz; above that the amplitude decreases sharply with frequency. Now, normal speech develops maximum average energy at about the same frequency and weakens at higher frequencies. A singer who produced sounds with the energy distribution of ordinary speech would therefore be in trouble: the orchestra's much stronger sounds would drown out the singer's. The average sound distribution of a trained singer, on the other hand, differs from that of normal speech—and of an orchestra—mainly because of the singing-formant effect. We have shown that a singer's voice is heard much more easily against recorded noise that has the same average energy distribution as an orchestra's sound if the voice has a singing formant. Not only is the formant almost invariably audible, because its frequency is in a region where the orchestra's sound is rather weak, but also it may help the listener to "imagine" he hears other parts of the singer's spectrum that are in fact drowned out by the orchestra.

The singing formant is at an optimal frequency, high enough to be in the region of declining orchestral-sound energy but not so high as to be beyond the range in which the singer can exercise

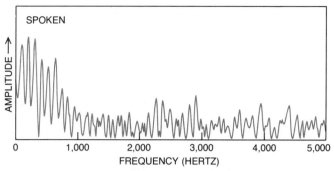

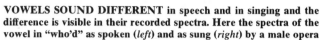

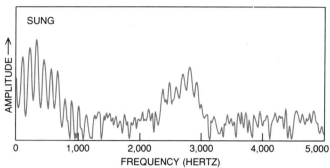

VOWELS SOUND DIFFERENT in speech and in singing and the difference is visible in their recorded spectra. Here the spectra of the vowel in "who'd" as spoken (*left*) and as sung (*right*) by a male opera singer are compared. What is significantly different about the sung spectrum is the spectral-energy peak that appears in it between about 2,500 and 3,000 hertz. The new peak is called the singing formant.

good control. Because it is generated by resonance effects alone, it calls for no extra vocal effort; the singer achieves audibility without having to generate extra air pressure. The singer does pay a price, however, since the darkened vowel sounds deviate considerably from what one hears in ordinary speech. In some kinds of singing that price is too high: the ideas and moods expressed in a "pop" singer's repertoire, for example, would probably not survive the deviations from naturalness that are required to generate the singing formant. And pop singers do not in fact darken their vowels; they depend on electronic amplification to be heard.

In cartoons a female opera singer is almost invariably depicted as a fat woman with her mouth opened very wide. In a study of female singers I have found that the way in which the jaw is manipulated is in fact quite different in ordinary speech and in singing. In speech the size of the jaw opening varies with the particular vowel, but in female singing it tends to depend also on the pitch of the tone that is being sung: the higher a soprano sings, the wider her jaw is opened. This suggested to me that a soprano must vary the frequency of her first formant according to the pitch at which she is singing. Analysis of formant frequencies confirmed that the articulation was being varied in such a way as to raise the first-formant frequency close to the frequency of the fundamental of the tone being sung. I noted such a frequency match whenever the frequency of the fundamental was higher than the frequency of a vowel's first formant in ordinary speech.

The reason becomes clear when one considers that the pitch frequency of a soprano's tones is often much higher than the normal frequency of the first formant in most vowels. If a soprano sang the vowel *ee* at the pitch of her middle C and with the articulation of ordinary speech, her first formant would be in the neighborhood of 270 hertz and the pitch frequency (the frequency of her lowest spectrum partial) would be almost an octave higher, at 523 hertz. Since a sound is attenuated in proportion to the distance of its frequency from a formant frequency, the fundamental would suffer a serious loss of amplitude. The fundamental is the strongest partial in the voice-source spectrum, and the higher its pitch is, the more important the fundamental is for the loudness of the tone, and so the singer's *ee* would be rather faint. Assume that her next sound was the *ah* sound of "father," to be sung at the pitch of high *F*. The fundamental, at 698 hertz, would be very close to the frequency of the first formant, about 700 hertz, and so the tone would be loud. The loudness of the singer's tones would vary, in other words, according to a rather unmusical determinant: the frequency distance between first formant and fundamental. In order to modulate the loudness accord-

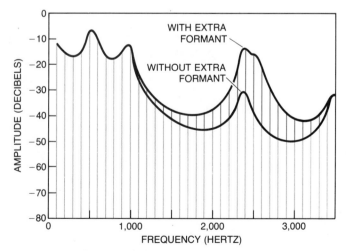

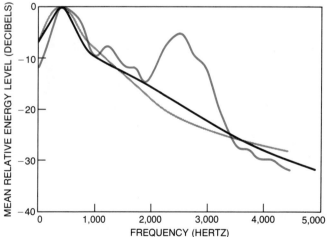

SINGING FORMANT'S ORIGIN (*left*) and its utility in singing (*right*) are demonstrated. An extra formant was inserted between the usual third and fourth formants in an experiment with an electronic resonator that behaves like the vocal tract (*left*). The new formant increased the amplitude of the partials near it by more than 20 decibels; similarly, an extra formant (achieved by lowering the larynx) supplies the high-frequency peak in the spectrum of a sung vowel. The three curves (*right*) show the averaged distribution of energy in the sound of orchestral music (*black*), of ordinary speech (*gray*) and of the late tenor Jussi Björling singing with an orchestra (*colored*). The distribution is very similar for speech and the orchestra at all frequencies; it is the singer's voice that produces the peak in the colored curve between 2,000 and 3,000 hertz. In that frequency region a singer's voice is loud enough, compared with an orchestra's sound, to be discerned.

ing to the musical context, the singer would need to continually vary her vocal effort. That would strain her vocal folds. (Experiments with synthesized vowel sounds suggest that it would also produce tones more characteristic of a mouse under severe stress than of an opera singer!)

The soprano's solution is to move the first formant up in frequency to match the frequency of the fundamental, thus allowing the formant always to enhance the amplitude of the fundamental. The result is that there is minimal variation in loudness from pitch to pitch and from vowel to vowel. Moreover, changing the size of the jaw opening in this way provides maximum loudness at the lowest possible cost in vocal effort. The strategy is probably resorted to not only by sopranos but also by other singers whose pitch range includes frequencies higher than those of the first formants of ordinary speech: contraltos, tenors and occasionally even baritones.

It can be hard for a student of singing to learn this special way of regulating the jaw opening, and particularly hard if the jaw muscles are under constant tension. That may explain why many singing teachers try to get their students to relax the jaw. Another frequent admoni-

tion is: "Hear the next tone within yourself before you start to sing it." That could be necessary because proper manipulation of the jaw opening requires some preplanning of articulation for particular vowels and for the pitch at which they are to be sung. Opening the jaw, however, is not the only way to raise the first-formant frequency. Shortening the vocal tract by drawing back the corners of the mouth serves the same purpose, and that may be why some teachers tell their students to smile when they sing high tones.

Since formant frequencies determine vowel quality, shifting the first-formant frequency arbitrarily according to pitch might be expected to produce a distorted vowel sound, even an unintelligible one. It does not have this effect, largely because we are accustomed to hearing vowels produced at various pitches in the ordinary speech of men, women and children with vocal tracts of very different lengths; if a vowel is high-pitched, we associate it with relatively high formant frequencies. The correlation is so well established in our perceptual system that we may perceive a change of vowel when we hear two sounds with identical formant frequen-

cies but different pitches; if a singer raises her first-formant frequency with the pitch, some of that rise is actually required just to maintain the identity of the vowel. It is true that when the pitch is very high, our ability to identify vowels deteriorates, but that seems to be the case no matter what the formant frequencies are. The soprano, in other words, does not sacrifice much vowel intelligibility specifically as a result of her pitch-dependent choice of first-formant frequency. (Incidentally, composers of vocal music are conscious of the problem of vowel identification at high pitches and generally avoid presenting important bits of text only at the top of a soprano's range; often the text is repeated so that the words can be well understood at a lower pitch.)

It is clear that a good deal of the difference between spoken and sung vowels can be explained by the singer's need for economy of vocal effort. The general idea is the same, whether in being heard over the orchestra or in maintaining loudness at high pitch: to take advantage of vocal-tract resonance characteristics so as to amplify sounds. The importance of these resonances, the formants, is paramount.

Confirmation of the importance of

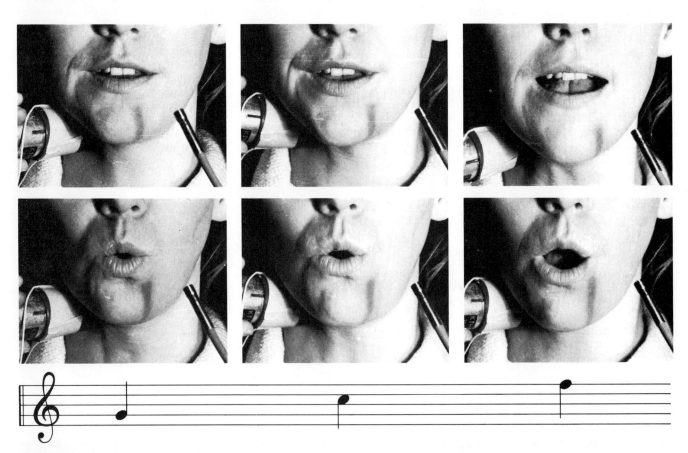

SOPRANOS and other singers of high tones tend to open their mouth wider with rising pitch. The tendency is demonstrated in these photographs of a soprano singing the vowel sounds of "heed" (*top*) and of "who'd" (*middle*) at successively higher pitches, shown in musical notation (*bottom*). When these photographs were made, the singer held a vibrator against her neck and a small microphone was placed near her lips. She began to sing each vowel at a specified pitch and then, with the vibrator turned on, she stopped singing but maintained the positions of the articulatory organs. The vibrator now supplied a steady, low-pitched sound that was influenced by the singer's vocal tract just as her own voice source would have been but that was more suitable for analysis than a high voice tone, which has few partials.

the formants was provided by a recent study of how male voices are classified as bass, baritone or tenor. Obviously the singer's frequency range is ultimately the determinant, but even when the true range (which is established primarily by the shape, size and musculature of the vocal folds) has not yet been developed, a good voice teacher can often predict the classification after listening to a student's voice. How is that possible? Thomas F. Cleveland, who was visiting our laboratory at the Royal Institute of Technology in Stockholm and is now at the University of Southern California, analyzed vowels sung by basses, baritones and tenors with respect to formant frequencies and the spectrum of the voice source. Then he had a jury of voice teachers listen to the vowel samples and classify the voices. The teachers tended to classify vowels in which the formant frequencies were comparatively low as having been sung by bass voices and vowels whose formant frequencies were high as having been sung by tenors. Variations in the voice-source spectrum (which varied slightly with the pitch at which a vowel was being sung), on the other hand, did not provide a basis for consistent classification. In a second test the same jury judged a series of synthesized (and therefore clearly defined) sounds and confirmed Cleveland's original impression: the lower the formant frequencies of a given vowel were, the lower the singer's voice range was assumed to be.

Cleveland found that typical bass and tenor voices differ in formant frequencies very much as male and female voices do. The formant-frequency differences between males and females are due mainly to vocal-tract length, and so the bass-tenor differences are probably also largely explained by the same physical fact. Formant frequencies are determined, however, not only by the individual's vocal-tract morphology but also by habits of articulation, which are highly variable. Be that as it may, vocal-tract morphology must set limits to the range of formant frequencies that are available to a singer.

At this point the reader who knows and cares about music may be rather disappointed. I have failed to mention a number of factors that are often cited as determinants of excellence in singing: the nasal cavity, head and chest resonances, breathing and so on. These factors have not been mentioned simply because they seem to be not relevant to the major acoustic properties of the vowel sounds produced in professional operatic singing. Our research suggests that professional quality can be achieved by means of a rather normal voice source and the resonances of the vocal tract.

Our implied model may not be perfect, to be sure. It is just possible, for

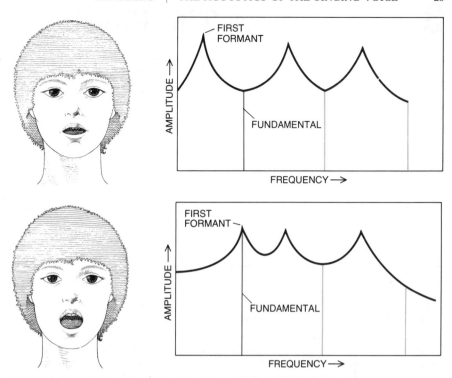

NEED FOR WIDE JAW OPENING arises from the fact that a soprano must often sing tones whose fundamental (lowest partial) is far higher in frequency than the normal first formant of the vowel being sung. When that is the case (*top*), the amplitude of the fundamental is not enhanced by the first formant and the sound is weak. Opening the jaw wider raises the pitch of the first formant. When the first-formant frequency is raised to match that of the fundamental (*bottom*), the formant enhances the amplitude of the fundamental and the sound is louder.

example, that the nasal cavity has a role in the singing of vowels that are normally not nasalized. If that is so, we have attributed its effect to the voice source, thus compensating for one error by making another. Moreover, we have dealt only with sustained vowel sounds, whose production is important but is certainly not the only acoustic event in singing.

Resonances outside the vocal tract, such as in the head or the chest, cannot contribute appreciably to the singer's acoustic output in view of the great extent to which sound is attenuated as it passes through tissues. This is not to say that such resonances may not be important to the singer, who may receive cues to his own performance not only from what he hears but also from felt vibrations. As for breathing, it is clear that the vocal folds would vibrate no matter by what technique an excess of air pressure is built up below the glottis. Breathing and laryngeal manipulation are likely to be physiologically interdependent, however, since the larynx is the gatekeeper of the lungs. Probably different ways of breathing are associated with different adjustments of the larynx, and probably some ways are effective for singing and others are inadequate or impractical.

Finally we return to the original question: What is so special about a singer's voice? The voice organ obeys the same acoustic laws in singing that it does in ordinary speech. The radiated sound can be explained by the properties of the voice-source spectrum and the formants in singing as in speech. From an acoustical point of view singers appear to be ordinary people. It is true that there is a major difference between the way formant frequencies are chosen in speech and the way they are chosen in singing, and hence between the way vowels are pronounced in singing and the way they are pronounced in speech. A man with a wide pharynx and with a larynx that will resonate at a frequency of between 2,500 and 3,000 hertz is likely to be able to develop a good singing voice more readily than a person who lacks those chacteristics. And his progress may be facilitated if his vocal folds give him a range that agrees with his formant frequencies. As for a female singer, she should be able to shift the first formant to join the pitch frequency in the upper part of her range; that requirement may bar some women with a long vocal tract from having a successful career as a coloratura soprano. There are, in other words, a few morphological specifications that probably have some effect on the ease with which someone can learn to sing well. There are other conditions that may be more important, however. It is in the complex of knowledge, talent and musical instinct that is summed up as "musicality," rather than in the anatomy of the lungs and the vocal tract, that an excellent singer's excellence lies.

3

The Physics
of the Piano

by E. Donnell Blackham
December 1965

Most musical instruments produce tones whose partial tones, or overtones, are harmonic: their frequencies are whole-number multiples of a fundamental frequency. The piano is an exception

Almost every musical tone, whether it is produced by a vibrating string, a vibrating column of air or any other vibrating system, consists of a fundamental tone and a number of the higher-pitched but generally fainter tones known as partial tones or overtones. The complex sound produced by this combination of separate tones has a timbre, or characteristic quality, that is determined largely by the number of partial tones and their relative loudness. It is timbre that enables one to distinguish between two musical tones that have the same pitch and the same loudness but are produced by two different musical instruments. A pure tone—one that consists solely of the fundamental tone—is rarely heard in music.

It is widely believed that the partial tones produced by all musical instruments are harmonic—that their frequencies are exact whole-number multiples of the frequency of a fundamental tone. This certainly holds true for all the woodwinds and under certain conditions for many of the stringed instruments, including the violin. It is only approximately true, however, in the most familiar stringed instrument: the piano. The higher the frequency of the partial tones of any note on the piano, the more they depart from a simple harmonic series. In our laboratory at Brigham Young University my colleagues and I, under the direction of Harvey Fletcher, have succeeded in measuring with considerable precision the degree to which the modern piano is inharmonic and have demonstrated the importance of this factor in determining the distinctive quality of the piano's tone.

The physics of the piano can best be understood by first reviewing the evolution of the modern piano and its principal components. Archaeological evidence shows that primitive stringed instruments existed before the beginning of recorded history. The Bible refers several times to an instrument called the psaltery that was played by plucking strings stretched across a box or gourd, which served as a resonator. A similar instrument existed in China some thousands of years before the Christian era. In the sixth century B.C. Pythagoras used a simple stringed instrument called the monochord in his investigation of the mathematical relations of musical tones. His monochord consisted of a single string stretched tightly across a wooden box. It was fitted with a movable bridge that could divide the string into various lengths, each of which could vibrate freely at a different fundamental frequency.

Another important component of the modern piano—the keyboard—did not arise in conjunction with a stringed instrument but with the pipe organ. The organ of Ctesibus, perfected at Alexandria in the second century B.C., undoubtedly had some kind of keyboard. The Roman architect Vitruvius, writing during the reign of Augustus Caesar, describes pivoted keys used in the organs of his day. In the second century A.D. Hero of Alexandria built an organ

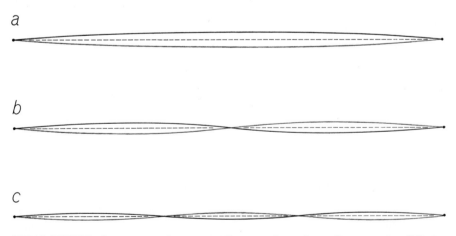

a

b

c

IDEAL STRING, that is, one without any stiffness, can be made to vibrate at many different frequencies: the fundamental frequency (*a*) produces a pure tone, rarely heard in music, whereas higher-pitched partial tones, or overtones, are produced by harmonic vibrations ("*b*" and "*c*"), whose frequencies are whole-number multiples of the fundamental frequency.

SIMULTANEOUS VIBRATION of a string at two or more different frequencies is normal for stringed instruments. Here the string vibrates at the fundamental frequency and the second partial frequency ("*a*" and "*c*" in upper illustration). In the piano the stiffness of the strings causes higher partials of a complex tone to depart from the simple harmonic series.

in which the valves admitting air to the pipes were controlled by pivoted keys that were returned to their original position by springs.

We do not know who first conceived the idea of adding keys to a stringed instrument. From this obscure beginning there eventually evolved in the 15th century the clavichord. In the early clavichords a piece of metal mounted vertically at the end of the key acted both as a bridge for determining the pitch of the string and as a percussive device for producing the tone [see upper illustration on page 28]. Since one string could be used to produce more than one tone, there were usually more keys than strings. A strip of cloth was interlaced among the strings at one end in order to damp the unwanted tone from the shorter part of the string.

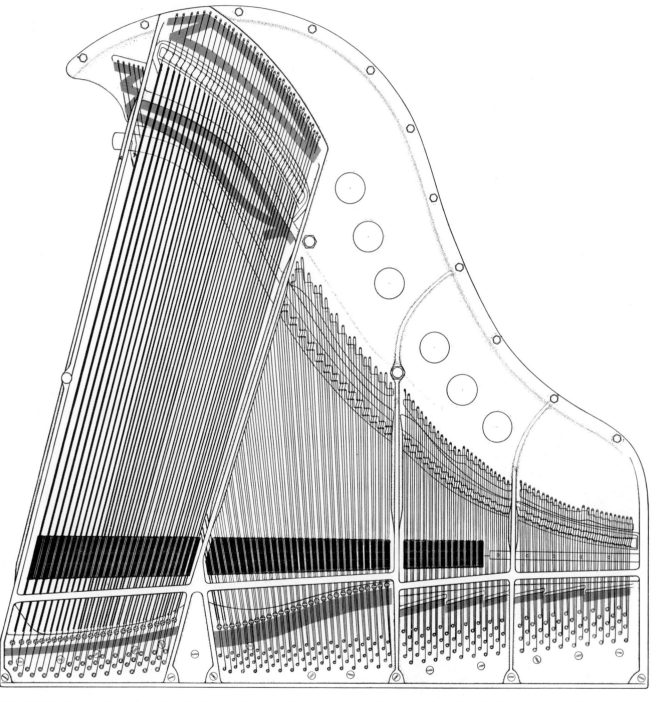

TOP VIEW of the interior of a modern "baby grand" piano shows the powerful construction of the full cast-iron frame, which sustains the tremendous tension exerted by the strings. In this particular piano the frame, which is cast in one piece, weighs about 250 pounds and sustains an average tension of some 50,000 pounds; in a larger concert-grand piano the frame weighs as much as 400 pounds and sustains an average tension of 60,000 pounds. The strings are made of steel wire with an ultimate tensile strength of from 300,000 to 400,000 pounds per square inch. In order to make the bass strings (left) vibrate slower and thus produce a lower pitch, they are wrapped in copper or iron wire; two such wrappings are often used in the extreme bass. In all modern pianos the bass strings are "overstrung" in order to conserve space and to bring them more nearly over the center of the soundboard. Starting from the treble, or right-hand, end of the keyboard there are 60 notes with three strings each, then 18 notes with two strings each and finally, in the extreme bass, 10 notes with only one string each. Larger pianos have more strings but the same total number of notes: 88. Rectangular black objects in a row near the bottom are the heads of the dampers. Parts made of felt are in color. Strips of cloth interlaced among the strings at top damp unwanted tones from the short parts of the strings beyond the bridge (see illustration on next page).

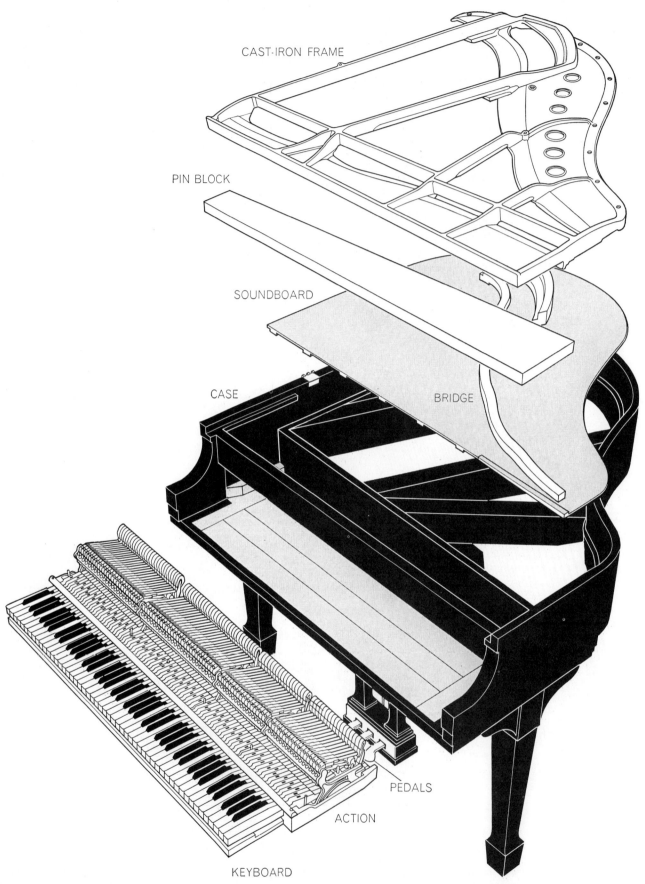

CAST-IRON FRAME

PIN BLOCK

SOUNDBOARD

CASE

BRIDGE

PEDALS

ACTION

KEYBOARD

EXPLODED VIEW of the baby-grand piano depicted from above on the preceding page shows the relations of several main components. The keyboard (*bottom left*) has 88 keys divided into seven and a third octaves. Each octave has eight white keys for playing the diatonic scale (whole notes) and five raised black keys for playing the chromatic scale (whole notes plus sharps and flats). Connected to the keyboard is the action, which includes all the moving parts involved in the actual striking of the string. Three pedals (*bottom center*) serve to control the dampers in the action. When a key is struck, the hammer sets the strings in vibration and, after a very short interval known as the attack time, sound is translated by means of a wooden bridge to the soundboard, from which it is radiated into the air. During the attack time sound is also radiated to a lesser degree from both the strings and the bridge.

Several essential characteristics of the modern piano are inherited from the clavichord. The clavichord had metal strings, a percussive device for setting the strings in vibration, a damping mechanism and also an independent soundboard: the board at the bottom of the case did not also serve as the frame for mounting the strings. Moreover, although the tone of the clavichord was weak, the instrument allowed for the execution of dynamics, that is, for playing either loudly or softly.

At about the same time another fore-runner of the modern piano was in process of development. In the spinet, or virginal, longer strings were introduced to produce a louder tone. Now the metal percussive device of the clavichord was no longer adequate to produce vibration in the strings. Instead the strings were set in motion by the plucking action of a quill mounted at right angles on a "jack" at the end of the key [see lower illustration on next page]. When the key was depressed, the jack moved upward and the quill plucked the string. When the jack dropped back, a piece of cloth attached to it damped the vibration of the string.

Around the beginning of the 16th century experiments with still longer strings and larger soundboards led to the development of the harpsichord. Although this instrument was essentially nothing more than an enlarged spinet, it incorporated several important innovations that have carried over to the modern piano. The wing-shaped case of the harpsichord is imitated by that of the grand piano. The stratagem of using more than one string per note in order to increase volume was adopted for the harpsichord by the middle of the 17th century. The harpsichord also had a "forte stop," which lifted the dampers from the strings to permit sustained tones, and a device for shifting the keyboard, both of which are preserved in the modern piano.

The invention of the piano was forecast by inherent defects in both the clavichord and the harpsichord. Neither the spinet nor the harpsichord was capable of offering the composer or performer an opportunity to execute dynamics. The clavichord, on the other hand, allowed a modest range of dynamics but could not generate a tone nearly as loud as that of the harpsichord. Attempts to install heavier strings in order to increase the volume of either instrument were futile; neither the metal percussive device of the clavi-

JURY composed of both musicians and nonmusicians was asked to distinguish between recordings of real piano tones and synthetic ones. When the synthetic tones were built up of harmonic partials, the musicians on the jury were able to distinguish 90 percent of these tones from real piano tones; the nonmusicians, 86 percent. When inharmonic partials were used, results showed that in most cases both the musicians and the nonmusicians were guessing; both groups identified only about 50 percent of the tones correctly. In this photograph two members of the jury are listening to tones in an anechoic, or echoless, chamber.

chord nor the quill of the harpsichord could excite a heavy string. Moreover, the cases of these early instruments were not strong enough to sustain the increased tension of heavier strings.

A remedy for these defects was provided by the Italian harpsichord-maker Bartolommeo Christofori, who in 1709 built the first hammer-action keyboard instrument. Christofori called his original instrument the "piano-forte," meaning that it could be played both softly and loudly. The idea of having the strings struck by hammers was probably suggested to him by the dulcimer, a stringed instrument played by hammers held in the hands of the performer. Christofori recognized that his new instrument would need a stronger case to withstand the increased tension of the heavier strings. By 1720 an improved

model of the pianoforte included an escapement device that "threw" the free-swinging hammer upward at the string and also a back-check that regulated the hammer's downward return [see upper illustration on page 29]. An individual damper connected to the action of the hammer was provided for each note.

For a century and a half after Christofori's first piano appeared inventors worked to improve the new instrument, particularly its novel hammer action. Several other types of action were developed, some new and others modeled closely on Christofori's original. Pianos were built in a variety of forms: traditional wing-shaped pianos, square pianos, upright pianos and even a piano-organ combination.

Among the major innovations toward the end of this period was the full cast-

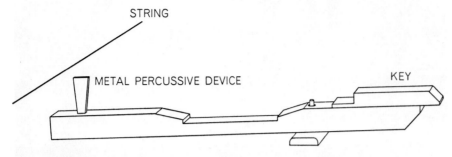

STRING

METAL PERCUSSIVE DEVICE   KEY

CLAVICHORD ACTION included one essential feature found in all modern pianos: a percussive device for setting the strings in vibration. A piece of metal mounted vertically at the end of the key acted both as a bridge for determining the pitch of the string and as a percussive device for producing the tone. Since one string could be used to produce more than one tone, there were usually more keys than strings. A strip of cloth was interlaced among the strings at one end to damp the tone from the shorter part of the string.

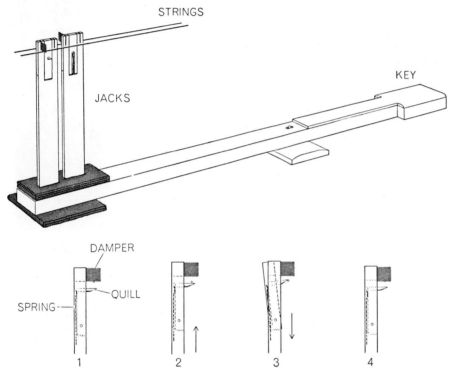

STRINGS

JACKS   KEY

DAMPER

QUILL

SPRING

1   2   3   4

HARPSICHORD ACTION was capable of producing a louder tone than that of the clavichord, but, unlike the clavichord, it did not allow for the execution of dynamics, that is, for playing either loudly or softly. The strings were set in motion by the plucking action of a quill mounted at right angles on a "jack" at the end of the key. When the key was depressed, the jack moved upward and the quill plucked the string. When the jack dropped back, a piece of cloth attached to it damped the vibration of the string. The stratagem of using more than one string per note was adopted in the harpsichord in the 17th century.

into the air. During the attack time sound is also radiated to a lesser degree from both the strings and the bridge. In the late 19th century Frederick Mathushek and his associates proved that the quality of a piano's sound was not influenced by the transverse, or horizontal, vibrations of the soundboard. They glued together two soundboards so that the grain of one was at right angles to the grain of the other, thereby eliminating any transverse vibrations, and found that the quality of the sound was not affected by this arrangement. The behavior of the soundboard has also been analyzed theoretically by a number of eminent physicists, including Hermann von Helmholtz, but such analyses have added nothing to the principles of soundboard construction arrived at empirically by the builders of the early clavichords and harpsichords.

The development of the full cast-iron frame gave the sound of the piano much greater brilliance and power. The modern frame is cast in one piece and carries the entire tension of the strings; in a large concert-grand piano the frame weighs 400 pounds and is subjected to an average tension of 60,000 pounds. In order to maintain the tension of the strings each string is attached at the keyboard end to a separate tuning pin, which passes down through a hole in the frame and is anchored in a strong wooden pin block. Since the piano would go out of tune immediately if the tuning pins were to yield to the tremendous tension of the strings, the pin block is built up of as many as 41 cross-grained layers of hardwood.

The keyboard of the modern piano is constructed on essentially the same principles that had been fully developed before the 15th century. The standard keyboard has 88 keys divided into seven and a third octaves, the first note in each octave having twice the frequency of the first note in the octave below it. Each octave has eight white keys for playing the diatonic scale (whole notes) and five raised black keys for playing the chromatic scale (whole notes plus sharps and flats). In all modern pianos the white keys are not tuned exactly to the diatonic scale but rather to the equally tempered scale, in which the octave is simply divided into 12 equal intervals.

The moving parts of the piano that are involved in the actual striking of the string are collectively called the action [*see lower illustration on opposite page*]. One contemporary piano manu-

iron frame. Constant striving for greater sonority had led to the use of very heavy strings, and the point was reached where the wooden frames of the earlier pianos could no longer withstand the tension. In 1855 the German-born American piano manufacturer Henry Steinway brought out a grand piano with a cast-iron frame that has served as a model for all subsequent piano frames. Although minor refinements are constantly being introduced, there have been no fundamental changes in the

design or construction of pianos since 1855.

A part of the piano that has received a great deal of attention from acoustical physicists is the soundboard. Some early investigators believed the sound of the piano originated entirely in the soundboard and not in the strings. We now know that the sound originates in the strings; after a very short interval, called the attack time, it is translated by means of a wooden bridge to the soundboard, from which it is radiated

facturer asserts that the action in one of his pianos has some 7,000 separate parts. Nearly all modern actions are versions of Christofori's original upward-striking ones, which took advantage of the downward force of gravity for the key's return. Some workers have experimented with downward-striking actions, so far without success.

Early in the history of piano-building the hammers were small blocks of wood covered with soft leather. The inability of leather to maintain its resiliency after many successive strikings led eventually to the use of felt-covered hammers. If the felt is too hard and produces a

harsh tone, it can be pricked with a needle to loosen its fibers and will produce a mellower tone. If the tone is too mellow and lacks brilliance, the felt can be filed and made harder.

A standard piano has three pedals that serve to control the dampers. The forte, or sustaining, pedal on the right disengages all the dampers so that the strings are free to vibrate until the pedal is released or the tones die away. The sostenuto pedal in the middle sustains only the tones that are played at the time the pedal is depressed; all the other tones are damped normally when their respective keys are released. The "soft"

pedal on the left shifts the entire action so that the hammers strike fewer than the usual number of strings, decreasing the loudness of the instrument.

The most interesting part of the piano from the standpoint of the acoustical physicist is of course the strings. The strings used in pianos today are made of steel wire with an ultimate tensile strength of from 300,000 to 400,000 pounds per square inch. Additional weight is needed to make the bass strings vibrate slower and so generate sounds of lower pitch; this is provided by wrapping the steel wire with wire of

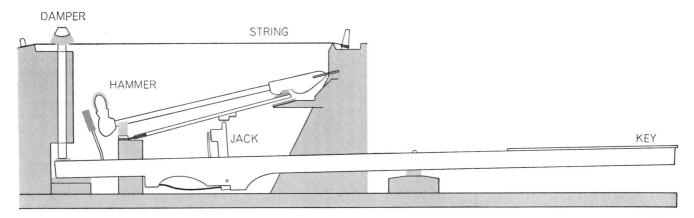

CHRISTOFORI ACTION, invented by Bartolommeo Christofori in the early 18th century, was the first hammer action and the prototype of all modern piano actions. It included an escapement device that "threw" the free-swinging hammer upward at the string

and also a back-check that regulated the hammer's downward return. An individual damper connected to the action of the hammer was provided for each note. Christofori called his instrument the "piano-forte," meaning it could be played either softly or loudly.

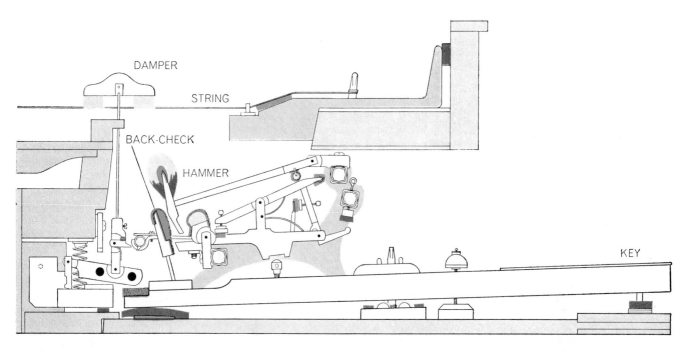

MODERN PIANO ACTION is modeled closely on Christofori's original upward-striking actions, which took advantage of the downward force of gravity for the key's return. Unlike the early hammers, which were small blocks of wood covered with soft

leather, the modern hammer is covered with felt. If the felt is too hard and produces a harsh tone, it can be pricked with a needle to loosen its fibers and will produce a mellower tone. If the tone is too mellow and lacks brilliance, the felt can be filed and made harder.

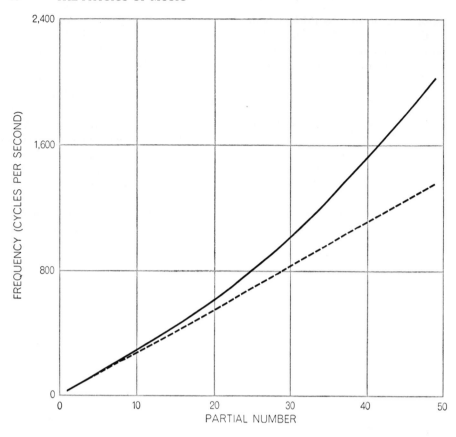

INHARMONICITY of a real piano tone is evident in this graph, based on data obtained from an electronic analysis of the partial tones of the lowest note on the piano keyboard (an *A*). The partials of the real piano tone (*solid line*) become increasingly sharper—that is, higher in frequency—compared with the partials of a pure harmonic tone (*broken line*).

copper or iron. Two such wrappings are often used in the extreme bass.

The vibration of a string that is attached securely at both ends is caused by a restoring force—a force that seeks to return the string to its original position after it has been displaced from that position. In a string that lacks stiffness the partial tones set up under the influence of the restoring force will be harmonic. In the piano the stiffness of the strings affects the restoring force to such a degree that some of the partials generated are not harmonic. This effect was known to Lord Rayleigh, who took it into account in formulating his classic equations for vibrating strings in the late 19th century. Many other investigators have since worked on the problem; our current effort is a continuation of the same line of inquiry.

Part of our program includes a series of tests in which a jury composed of both musicians and nonmusicians is asked to distinguish between recordings of real piano tones and synthetic ones. The synthetic tones are made by a bank of 100 audio-frequency oscillators that can be tuned to cover a range of from

50 to 15,000 cycles per second. Fine tuning is achieved by means of an attenuator connected to each oscillator circuit; the attenuator covers a range of 50 decibels, 10 decibels being a tenfold increase or decrease in the intensity of sound. With this apparatus it is possible to build up synthetic tones that represent a wide variety of partial-tone combinations. Real piano tones can be closely imitated by tuning a separate oscillator to the precise frequency and intensity associated with each partial tone of the real tone. The complex synthetic tone thus generated can then be fed into an "attack and decay" amplifier in order to give it the attack-and-decay characteristics found in the real piano tone.

In our early tests the synthetic tones were arbitrarily built up of harmonic partials. The musicians on the jury were able to distinguish 90 percent of these tones from real piano tones; the nonmusicians, 86 percent. In later tests synthetic tones built up of inharmonic partials were used. Results from these tests showed that in most cases both the musicians and the nonmusicians were guessing; both groups identified only about 50 percent of the tones correctly.

Whenever a synthetic tone and a real tone were judged to be identical, we could give a description of the quality of the real tone based on our knowledge of the quality of the synthetic tone.

Recorded piano tones can also be analyzed directly by means of a conventional audio-frequency analyzer that is adjusted to pass only a narrow band of frequencies (about four cycles per second). The analyzer is set at different frequencies until it registers a maximum response for the particular partial being analyzed. A pure tone from one of the oscillators is then sent through the analyzer, and its frequency is adjusted until it gives a maximum response at the same setting as that of the real partial. An electronic counter is used to measure the frequency of the oscillator tone to an accuracy of within about a tenth of 1 percent. This frequency is assumed to be the frequency of the real partial being analyzed.

A sample of this kind of analysis for the lowest note on the piano keyboard (an *A*) is given in the illustration at the left. It is evident that the partials of the real piano tone become sharper— that is, higher in frequency—compared with the partials of a pure harmonic tone. The 16th partial, for example, is a semitone sharper—half a step higher— than it would be if it were harmonic. The 23rd partial is more than a whole tone sharp, the 33rd partial is more than two tones sharp and the highest partial in the analysis, the 49th, is 3.65 tones sharp.

In addition to the fact that the piano's tones are generally inharmonic, the partials of any particular note tend to vary considerably in loudness. This variation is called the harmonic structure of the tone, or in the case of the piano, the partial structure. One way to analyze the partial structure of a piano tone is to measure the maximum response of each partial as it passes through the audio-frequency analyzer. This method was used to obtain the partial structure of the four *G*'s shown in the illustration on the opposite page.

The foregoing method does not yield the most accurate description of the partial structure of a piano tone, because the structure is continuously changing. When a piano string is struck by its hammer, its response reaches a maximum an instant later. From this moment on the tone dies away as the string gradually ceases to vibrate. Because the ear perceives the entire tone dying away uniformly, it might seem that all the partials of the tone die away at an equal rate. An examination of the decay curves of individual partials proves that this is not the case [*see illustration on next page*]. It is obvious from these curves that if the partial structure of a tone were measured at any given time, it would be different from the structure at any other time. Nonetheless, some authors still refer to a decay rate of a tone as so many decibels per second. In actuality the partials do not all decay at the same rate; in some cases they may even increase in intensity before starting to decay.

The tones used for our decay-time analyses were recorded in an ordinary music studio. It was thought at first that the irregular variations during decay might be related to the acoustic characteristics of the room or the piano. Accordingly the experiment was repeated in three different rooms: a normally reverberant studio, a very reverberant room and an anechoic, or echoless, chamber. The irregularities in the decay curves were present in all three rooms [*see illustration on page 33*].

One of the main advantages of our synthetic-tone system is that it can be used to produce synthetic tones identical with one another and with a real tone except for certain selected characteristics. For example, a group of synthetic tones can be produced that differ only in attack time, the time required for the loudness of the tone to reach its first maximum after the hammer strikes the string. By presenting such a group of tones to our jury we were able to determine that for the *G* just above middle *C* the attack time has to be between zero and .05 second to sound like the *G* on a piano. An attack time in the range of from .05 to .12 second made the note seem questionable, and one longer than .12 second made it sound decidedly unlike a *G* struck on a piano. For lower notes the required attack time tended to be longer; for higher notes it tended to be shorter.

Synthetic tones can also be produced that are identical with one another and with a real tone in every respect except decay time, the time required for the string to stop vibrating after it has reached its maximum loudness. For an undamped *G* above middle *C* the decay time required for the synthetic tone to sound piano-like was between two and 5.5 seconds. Again acceptable decay times were longer for lower notes and shorter for higher notes.

Another procedure is to give synthetic tones a piano-like attack and decay but

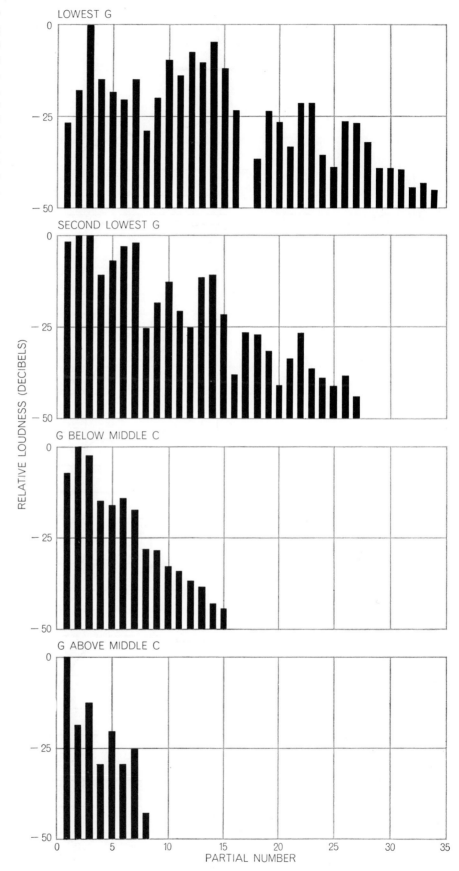

PARTIAL STRUCTURES of the four lowest *G*'s on the piano keyboard are presented in these four bar charts. The partial structure of a musical tone is the variation in loudness of the partial tones that constitute that particular tone. The partial structures of these four notes were obtained by measuring the maximum response of each partial as it passed through an audio-frequency analyzer that was adjusted to pass only a narrow band of frequencies. The readings are given in relative decibel levels with the loudest partial of each note set at zero; the other partials can then be read as so many decibels below zero.

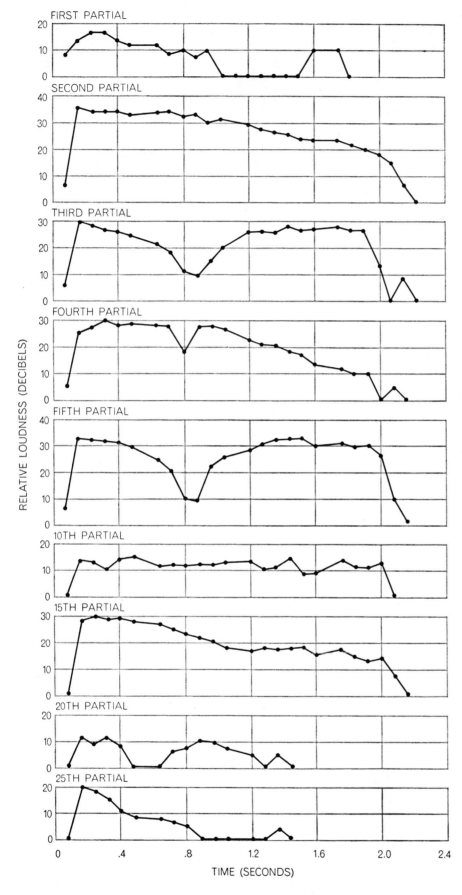

FIRST PARTIAL

SECOND PARTIAL

THIRD PARTIAL

FOURTH PARTIAL

FIFTH PARTIAL

10TH PARTIAL

15TH PARTIAL

20TH PARTIAL

25TH PARTIAL

RELATIVE LOUDNESS (DECIBELS)

TIME (SECONDS)

DECAY CURVES for nine partial tones of the lowest *C* on the keyboard demonstrate that the partial tones of a piano note do not all die away from an initial maximum at the same rate. In some cases they may even increase in loudness before beginning to decay. For each curve 30 measurements were made at equal intervals of .08 second each. Obviously the partial structure of a tone at any given time is different from the structure at any other time.

to vary the partial structure. In one test synthetic tones were built up in such a way that the loudness of each successive partial was a constant number of decibels less than that of the partial just below it in frequency. For example, if the difference was two decibels, then the second partial would be two decibels fainter than the first partial, the third partial would be four decibels fainter than the first, and so on. The limits of this "decibel difference" for obtaining a piano-like tone from the *G* above middle *C* were from five to 13 decibels per partial. In this case the acceptable range was narrower for lower notes and wider for higher notes. Tones produced when the decibel difference was below the lower limit were judged by the jury to sound "dead" or "hollow." Tones above the upper limit were described as sounding "like a harpsichord" or having "too much edge."

Synthetic tones that were built up of perfectly harmonic partials were described by the musicians and nonmusicians alike as lacking "warmth." Musicians generally use this term to suggest a certain quality of musical tone. For instance, a number of violins playing the same note at the same time produce a tone that is said to be warmer than that produced by a single violin playing alone. This quality appears to result from the fact that it is impossible for a number of musicians to play exactly in tune. When two different frequencies are sounded together, "beats" can be detected, the number of beats being equal to the difference in cycles per second between the two tones. A difference as small as two cycles per second between the fundamental frequencies of two tones can, however, produce much larger differences in the upper partials. Thus the beats that occur when two tones, each with a large number of partials, are sounded simultaneously can be quite complex. It is such beats between tones that account for the warmth produced by several violins or by a chord on the piano.

In the piano some additional warmth can be attributed to the fact that most of the hammers strike more than one string at a time. If the strings are not identically tuned, beats will occur between the high partials produced by each string. If such beats become too prominent, of course, the strings are declared to be out of tune.

The quality of a piano's tone also depends on several outside influences that

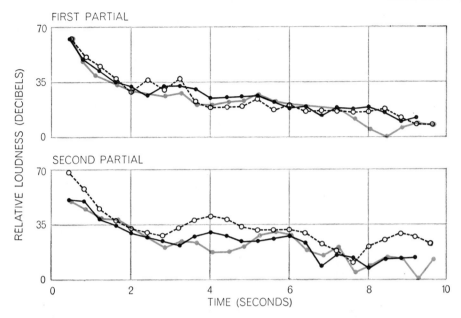

FIRST PARTIAL

SECOND PARTIAL

TIME (SECONDS)

RELATIVE LOUDNESS (DECIBELS)

ACOUSTICS OF ROOM in which the tones used in the decay-time analyses were recorded were shown by the author and his colleagues to have a negligible effect on the irregularities present in the decay curves of different partial tones of the same note. To obtain these curves the decay times for the first and second partials of the *G* above middle *C* were recorded in three different rooms: a normally reverberant studio (*broken black curves*), a very reverberant room (*solid black curves*) and an anechoic chamber (*solid colored curves*).

are not usually considered intrinsic properties of a vibrating string. There is the impact noise of the hammer as it strikes the string, the mechanical noise of the damping pedals, the effect of the damper on the end of a tone, and the noise level of all the other strings, which are free to vibrate sympathetically when they are not damped. In early tests it became quite evident that our juries were using these factors as clues to distinguish the real tones from the synthetic ones.

The impact noise of the hammer is not as noticeable in the lower register as it is in the upper. For the high strings the impact noise is almost as loud as the tone itself. A similar noise had to be superposed on the synthetic tones before they could be effectively used in our tests. Preference tests were set up to see if piano tones without this noise were more acceptable musically than tones in which the noise was present. In general the individuals tested were satisfied with the quality of piano tones as it is, and any large departures from this quality were disparaged. Obviously this is the result of years of conditioning, of hearing piano tones pro-

duced by pianos. Some composers even write with this specific quality in mind. An example can be found in *Piano Concerto No. 2* of the American composer Edward MacDowell, in which certain passages are marked *martellato,* presumably to indicate that as much hammer noise as possible should be introduced into the passage.

The mechanical action of the pedals or dampers also makes a noise that has become part of the piano's tone. Moreover, there is a distinctive effect evident when the felt on the dampers is brought into contact with the string: the tone is not cut off cleanly but rather sounds as though it is being swallowed. The problems involved in trying to duplicate these "side effect" sounds can be eliminated by using piano tones that are produced by striking a key and allowing the sound to decay naturally by holding the key down. In this way all the other strings remain damped. The pedals are not used and only the damper of the struck string is disengaged by the action of the key.

Our studies have clearly shown that a complete description of the quality of the piano's tone must contain more than partial structure, attack time and decay time. Above all, the inharmonicity of the piano's tones must not be neglected. Some believe that the tone quality of the piano could be improved merely by making the tones more harmonic. Our tests have proved that synthetic tones built of harmonic partials lack the quality of warmth that is associated with the piano as it exists today.

4

The Physics
of Wood Winds

by Arthur H. Benade
October 1960

*For centuries craftsmen have fashioned these
instruments by trial and error and rules of thumb. The
principles that underlie these empirical methods are
now being studied by the modern physicist*

There are, of course, hundreds of directions in which the mind of a physicist can stray while he sits in a concert hall, but one in particular unites his esthetic and scientific interests in a most fruitful manner. He can, if he likes, trace the history of much of his discipline in terms of the study of music and the instruments that make it. As far back as Pythagoras it was recognized that the most prominent pitch intervals are obtained by shortening a harp string so that some simple fraction (1/2, 1/3 and so on) of its length is left free to vibrate. Study of the vibration of strings by René Descartes' collaborator Marin Mersenne laid a foundation for the study of partial differential equations and their applications, a development whose roots are found in the work of such great mathematical physicists as Daniel Bernoulli and Jean le Rond d'Alembert in the 18th century. In the 19th century Hermann von Helmholtz devoted a large fraction of his enormous talents to the study of vibrating systems, as did Lord Rayleigh (who died in the 20th century, leaving his *Theory of Sound,* which is still a classic).

The rise of quantum mechanics has made the understanding of the underlying classical wave-theory even more important to physicists than it was in the days of Rayleigh. As a result every student of physics, whether he is interested in music or not, is faced daily with classroom problems that are variations on a theme born originally of music.

Since the mid-1920's, however, the engrossing new questions of quantum physics have diverted the energy of both theoreticians and experimenters from the more traditional lines of study, and so brought active musical research largely to an end. Still, the stage is set for a revival of musical physics. The techniques of measurement and calculation that have developed during the last 40 years in other areas of physics may now make it possible to solve problems in music that have withstood the best efforts of the past.

Some of the most stubborn questions are posed by the search for an orderly connection between the physical properties of instruments and the musical sounds that issue from them. The instruments are all of ancient descent, antedating Pythagoras and others who first felt the strong pull of music on the scientific as well as on the artistic imagination. The various winds and strings assumed their modern forms by a process akin to biological evolution. Trial and error, rule of thumb and traditions handed down from generation to generation of instrument makers account for the characteristic quality of the sound that each one generates, as well as for their peculiarly individual appearance and design. Yet much of the anatomy, and therefore of the performance, of an instrument is susceptible to analysis. One is led to ask if it is possible to create, with the help of theoretical considerations, new ways of constructing instruments whose performance is the same as those that have grown from empirical invention. Can we clarify the problems facing players and builders of these instruments in a way that suggests answers to hitherto unsolved problems? Can we invent entirely new musically useful instruments?

My own interest in these questions began with the playing of wood winds—

the tubular horns that bristle with levers, buttons and rings, and are played by the musicians who sit in the middle background of the symphony orchestra. Most familiar are the clarinet, the oboe, the bassoon, the saxophone, the English horn and the flutes (including the recorder, the fife and the piccolo). But the family also includes the bagpipe and such less well-known instruments as the arghool and the chalumeau (relatives of the clarinet), the aulos, or Greek flute, and the shawm (a relative of the oboe).

All wood winds may be disassembled mentally into three essential parts: the reed, the bore and the side holes. Air blown into the instrument through the reed sets up vibrations in the column of air within the bore, and this vibrating air column produces the sound of the instrument. The frequency at which the air vibrates is determined chiefly by the dimensions of the bore. These dimensions are modified in turn by the side holes in both their open and closed positions, as will be seen.

The reed system acts as a valve. It replenishes the vibrational energy of the air in the bore by converting a steady flow of compressed air from the player's lungs into a series of puffs at the frequency dictated by the bore. This valving of the air supply is accomplished differently in the flutes than in the other wood winds, but the device that does it may nevertheless be called a reed. The reed valves of all wood winds except the flutes are pressure-operated. They consist either of a single blade of cane fitted to a mouthpiece (as on the clarinet and saxophone) or a double blade of cane (oboe and bassoon). Vibration of the reed opens or shuts the thin slit (between the reed and the mouthpiece or between the two reeds) through which

FAMILIAR WOOD WINDS (*from left to right*) include the bassoon, oboe, flute, clarinet, English horn and saxophone. The wood-wind family also includes the bagpipe, the piccolo and several other instruments.

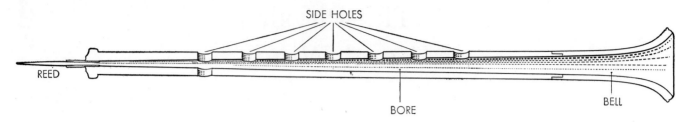

MAIN PARTS of a wood wind are the reed, the bore and the side holes. Oboe shown in this schematic cross section has fewer holes than actual instrument. Oboes, English horns and clarinets are equipped with bells; flutes, saxophones and bassoons are not.

the air is blown into the bore. The frequency of vibration is set by the cyclic changes in the pressure of the vibrating air in the bore. The mass and stiffness of the cane give the bore almost complete domination over the reed in determining the pitch. This domination of the reed by the bore is one of the distinguishing characteristics of wood winds. In contrast, the tube, or bore,

of a brass instrument, such as the trumpet or trombone, strongly influences the "reed" (the vibrating lips of the player) but does not dominate it; the same is true of the relation between the pipes and reeds used in pipe organs.

In flutes the function of the reed is served by a thin jet, or "reed," of air blown across the mouth hole in the side of the instrument. The vibrations of an

air-jet reed are controlled by cyclic changes not in the pressure but in the velocity of the air at the upper end of the bore. Rushing in and out of the bore at right angles to the reed, the vibrating air column drives the reed-jet upward and downward at the frequency fixed by the bore and side holes [*see illustration on pages 38 and 39*].

Because cane reeds are pressure-oper-

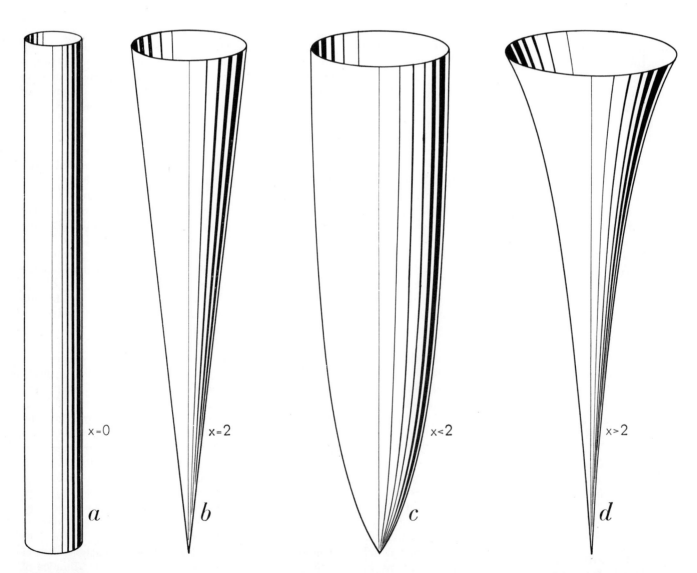

BESSEL HORNS increase in cross-sectional area according to some exponent (x) of the distance from the end of the horn. Bores of all practical wood winds are either cylindrical Bessel horns (a) or conical Bessel horns (b). Horns c and d are musically useless.

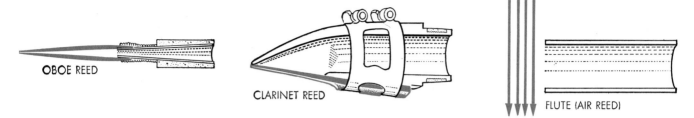

OBOE REED

CLARINET REED

FLUTE (AIR REED)

REEDS (*color*) act as valves that control flow of air into a wood wind. Cane reeds of oboe and clarinet are opened and shut by changes in pressure of the vibrating air in the bore; air-jet "reed" of flute is controlled by changes in velocity of the air in the bore.

ated, they will function only when the vibrations of the air column in the bore are such as to produce maximum variation in pressure at the mouthpiece end of the bore. The air-jet reed of the flute correspondingly requires vibrations that produce a maximum variation in velocity at the same end of the bore. Since the end of the bore away from the player's mouth is essentially open to the atmos-

phere, the variation in air pressure at that end can only be very small, and, with the air free to flow in and out, the variations in velocity will consequently be large. In short, the operation of a cane reed calls for those vibrations in the air column that produce maximum variation of pressure at the reed end of the bore, and essentially zero fluctuations at the lower end, while an air-jet

reed will sustain only those vibrations that yield a maximum fluctuation in velocity at both ends. It is interesting that this specification of the "end conditions" for a bore is in precisely the same form as that which mathematicians use in solving problems of vibrating systems involving no reed at all. Theory and experiment agree that only a certain discrete set of vibrational frequen-

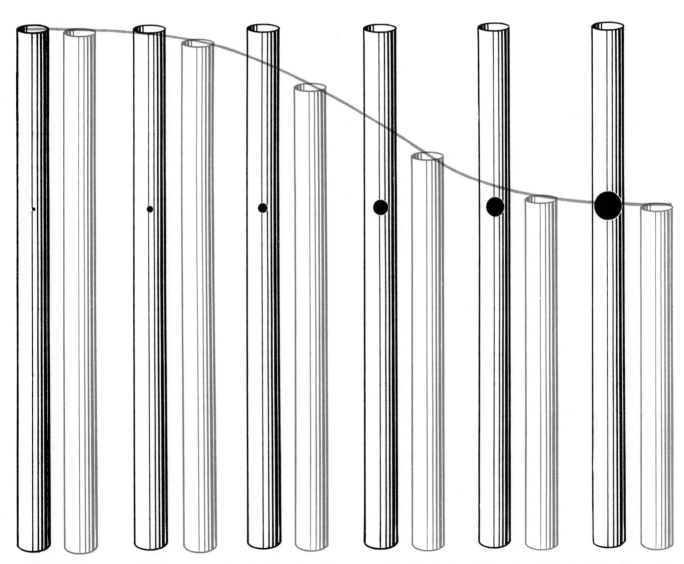

SIDE HOLES decrease the effective length of a bore. In this schematic diagram full-length bore is shown in black; its corre- sponding effective length when pierced by side holes of increasing size is shown in color. Holes permit musician to play a scale.

cies is possible for the air within a bore when it moves in a way that meets these conditions. These types of vibration (which are mathematically similar to the quantized states of motion of an electron in an atom) are called the "normal" or "natural" modes of vibration of the air in the bore.

The frequency and wavelength of these natural modes of vibration depend chiefly on the length and shape of the bore. For example, a cylindrical bore open at one end and blown at the other with a velocity-controlled air-jet reed (like that of a flute), or a conical bore blown at the small end with a pressure-controlled cane reed (as in the oboe), has a lowest, or fundamental, frequency of vibration whose sound has a wavelength twice the length of the bore; it can also vibrate in higher modes producing a sequence of notes that is often referred to as the harmonic series, with wavelengths that are integral fractions (1/2, 1/3, 1/4, 1/5 and so on) of the fundamental. A cylindrical bore blown by a pressure-controlled reed at one end vibrates in its lowest mode to produce a fundamental note whose wavelength is four times the length of the bore. The higher-frequency modes of vibration are those whose sound wavelengths are odd-numbered fractions (1/3, 1/5, 1/7 and so on) of the fundamental. Wood winds make extensive use of only the lowest three or four of their natural modes; the lowest mode corresponds to the so-called low register of the instrument, the second mode to the middle register, while the upper register uses one or another of the higher modes.

When a bugler plays reveille, he uses his lips as a pressure-controlled reed to excite one or another of the natural modes of vibration of a flaring brass pipe. It is the task of the bugle manufacturer to shape the instrument so that the frequencies of its natural modes are those desired for playing music. As everyone knows, the repertory of the bugle is rather limited: even a skilled bugler can sound only five or six widely spaced notes.

In contrast, wood winds can play a chromatic scale of more than 30 notes, because their makers have found ways to fill the gaps in pitch between the natural frequencies of the bore. This is accomplished by drilling a row of holes in the side of the bore. As intuition suggests, a hole just a few thousandths of an inch in diameter will cause little change in the pitch of a note played on

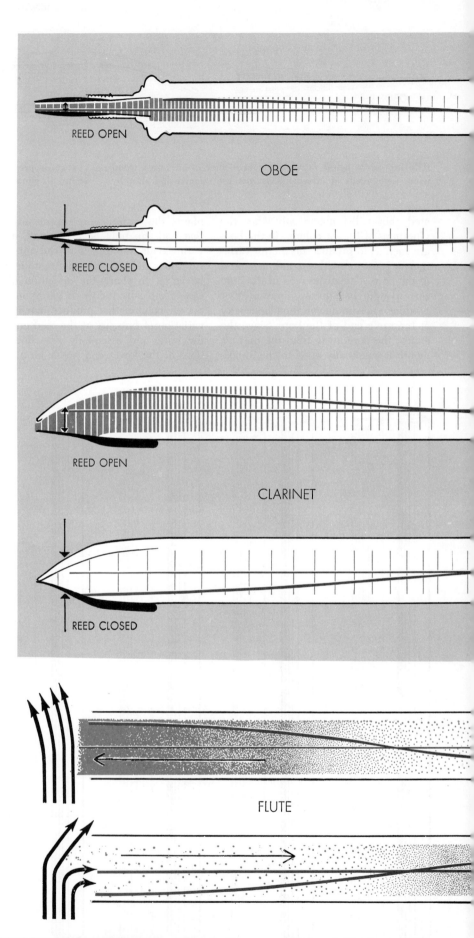

BORE DOMINATES REED in all wood winds. Pressure-operated reeds like those of oboe (*top*) and clarinet (*middle*) function only at frequencies at which bore produces maximum variation in pressure at the reed. Velocity-operated reed of flute (*bottom*) operates only at frequencies at which bore produces a maximum variation in velocity at the reed. Heavy

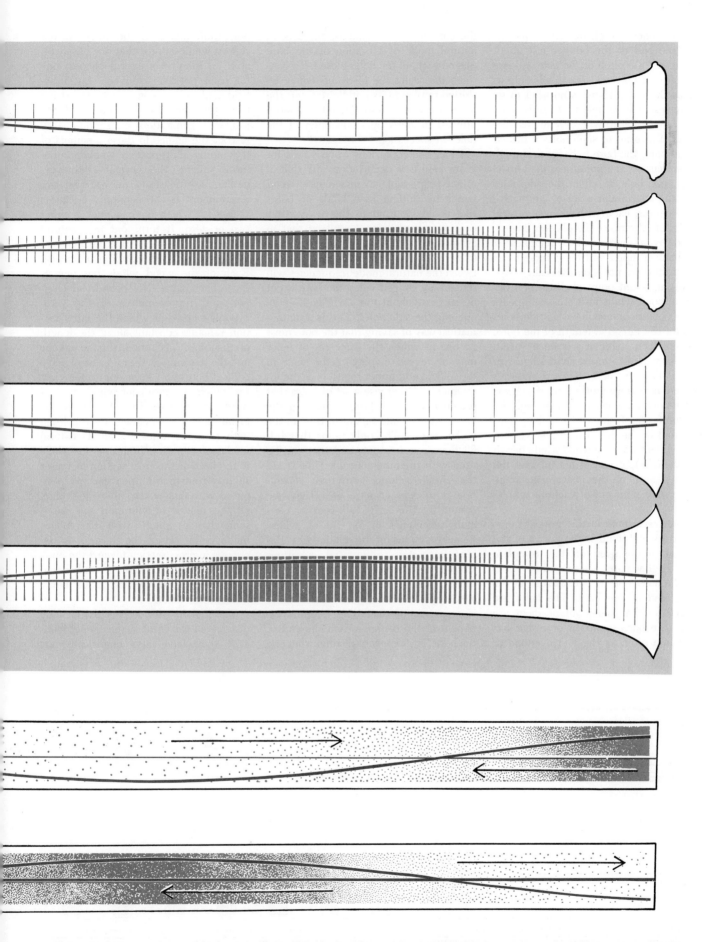

black lines represent reeds; gray lines show wave motion of vibrating air in bore. Heavy colored lines indicate high-pressure phases; lighter lines, low-pressure phases. In the flute heavy color indicates high-velocity motion in one direction; light color, motion in the opposite direction. Heavy black arrows show the vibrations of the air-jet reed. The instruments in this schematic diagram are shown playing in their second mode of vibration. The thickness of the bores and the relative size of the reeds are exaggerated.

the instrument. But if the hole is so large that the end of the bore is just about falling off, the pitch of the note will rise to that of a bore which extends only as far as the hole. There is no need to go to this extreme; a hole that can be covered by a fingertip will serve to raise the pitch. If a series of such holes is drilled in the bore, the bore behaves as though it were cut off at a point near the uppermost open hole. In effect the side holes give the instrument a set of alternative bore-lengths, each with its own natural modes of vibration. The player's fingers, aided by a more or less complicated mechanism, open and close these holes to get different notes on the scale.

An important point of distinction between wood winds and brasses appears at this point: brasses have no side holes to alter the effective length of the bore. Brass players achieve a scale by using a set of valves to insert short lengths of tubing into the bore. Thus if water is poured into the mouthpiece of a brass instrument, all of it will flow through the convolutions of the coiled tubing and pour out of the bell. But if one were to pour water through a wood wind (perish the thought), it would not all flow the whole length of the instrument; some of it would pour out of each open hole.

The lowest note in the musical range of a wood wind is the lowest of the natural frequencies of the complete bore. Higher notes are played by opening the holes one by one, starting at the lower end of the bore. The holes are spaced so that when all of them are open, the bore sounds a note that has the same pitch as the first overtone of the complete bore, that is, the second natural mode of vibration of the bore played with all the holes closed. To continue the scales the instrument may be shifted to play in its middle register, the holes are closed and the pitch altered upward in steps by again opening the holes in succession.

I had played various wood winds for years before it occurred to me that it might be possible to invent new wood winds by finding new kinds of bores that could be used in this way. After considerable thought I realized that in essence the problem is to find a class of horn shapes for which the ratios between the natural-mode frequencies will remain unchanged when the bore is cut off successively at the lower end, as by opening the side holes. This is a mathematical way of saying that the bore must be one in which the same set of holes may be reused in playing the scale in the middle and upper registers of the instrument. (Otherwise a set of holes that provided notes of proper pitch in the low register would be out of tune in the higher ones.) The only horns that fulfill these requirements and that can be used with pressure-controlled reeds are the so-called Bessel horns [*see illustration at bottom of page 36*]. They are named for the early 19th-century German astronomer F. W. Bessel, whose Bessel functions for the relation of variables in certain differential equations have been utilized in many areas of physics. One of these equations applies to waves in a series of horns that increase in cross section according to some exponent of the distance from the end of the bore. The exponent distinguishes one horn from another in the series. At first sight it would appear that one has available an infinity of useful shapes, one for each positive exponent. But practical and subtle considerations arising from the nature of our ears and the proper regeneration of sound-energy restrict the choice to those members of the family that possess whole-number ratios among the normal frequency-modes. A little study showed that this requirement is satisfied only by those shapes for which the exponent is either two or zero. It turns out that natural selection in musical instruments had long anticipated this finding of theory: Each of the wood winds has one or the other shape. The cylindrical clarinet is a representative of the zero-exponent class, in which the cross section remains constant along the length of the bore. The oboe, the saxophone and the bassoon, on the other hand, have conical bores and so belong to the class for which the exponent equals two. A bore that departs from these "ideal" forms does not enclose an air column with constant ratios among its natural modes of vibration from one open hole to the next. If the cross section increases slightly toward the open end, as compared with one or the other ideal bore, all the modes of vibration are raised somewhat in pitch, with the lowest mode being raised the most; if it decreases in cross section toward the open end, the modes of vibration are lowered in pitch, with the lowest mode being lowered the most.

The makers of wood winds not only discovered the ideal shapes generations ago; they have also empirically ex-

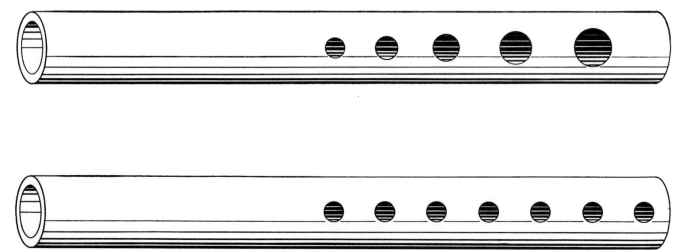

OPEN HOLES of an actual wood wind (*top*) increase in size and spacing toward the open end of the bore. But the frequencies at which the air within the bore vibrates can be calculated by assuming that the holes are all of uniform size and spacing (*bottom*).

ploited the effect of departures from the ideal. They judiciously alter the cross section of the bore by a few thousandths of an inch to compensate for various upsetting effects caused by the complex behavior of reeds and holes. As a result the bores of actual wood winds are not perfect cones or cylinders. Although the necessary modifications of a bore can in general be predicted quantitatively by proper mathematical analysis, to my knowledge such methods have almost never been employed by the manufacturers of wood winds. They make the final adjustments in the taper of each bore by a process of trial and error.

The size and position of the side holes are just as crucial as the shape of the bore in affecting the performance of an instrument. Not only do the side holes cut off the bore at a convenient spot for getting a scale; they also play a large role in setting the tone quality of the sound within the instrument. And when they are open, they influence the way in which this sound is ultimately radiated into the air for the listener to hear. A length of plastic tubing played with a clarinet mouthpiece gives a dull, plumbing kind of sound that few people can identify. When this same pipe is provided with a row of closed side-holes followed below by four or five open ones, the tone changes strikingly into the woody voice of a clarinet.

The "unused" closed holes convert the bore from a smooth-walled pipe into a lumpy duct that may be looked upon as a pipe with a series of swellings, as shown in the illustration below. Using

a mathematical method devised by Lord Rayleigh, I was able to calculate that if the musical properties needed by the bore are to be preserved when it is supplied with closed finger holes, then the size of the holes must be related to their spacing in a certain definite way. One can say that the ratio of the volume of air contained in the closed hole divided by the volume of air in the length of bore between adjacent holes must be the same in all parts of the bore. I was able to verify this deduction from theory by a quick measurement of the hole sizes on present-day instruments. In all pressure-controlled reed instruments the holes must be larger and farther apart toward the lower end of the bore. This does not apply to the velocity-controlled flute family, which is much less sensitive to such perturbing effects, so that all the holes can be roughly the same size.

For generations craftsmen have used rules of thumb to determine where the holes should be placed in a wood wind. But a given set of rules applies only to a particular design, and must often be adjusted to correct vagaries in tuning between registers. The ability to make first-class oboes, clarinets and flutes is often a matter of highly prized family craftsmanship. Precision mass-production techniques must often be supplemented by painstaking handwork to maintain any sort of quality. Theoretically, of course, the positioning of the holes is determined by the physics of the musical scale, whether we understand it or not. Around 1930 the late E. G. Richardson of University Col-

lege London, using electrical-analog techniques devised in 1919 by the late A. G. Webster of Clark University, calculated the behavior of a single open hole in the side of a tubular bore. Although this technique yielded a precise result and could in principle be extended to describe the musical case where there are several open holes, in practice it is too cumbersome to be usable. Since manufacturers of instruments were already able to make excellent wood winds without theoretical help, Richardson's work has lain largely neglected.

The difficulty with Richardson's approach is that each hole must first be treated separately, and then the mutual effects of the holes must be reconciled. A more practical approach is to study several adjacent holes simultaneously in a simplified way. A method for doing this in electrical systems was devised in 1927 by W. P. Mason of the Bell Telephone Laboratories. He worked out a set of equations to show the effect of regularly spaced loading coils on the vibrations in electrical transmission-lines. Because of the similarity of all vibrating systems, the equations can also be used for describing the sound vibrations in a bore with regularly spaced discontinuities such as holes. At first glance this does not seem useful because the holes along a wood-wind bore are not evenly spaced. But as all wood-wind players know, when their instrument plays a note in one of its lower two registers, only the two or three nearest open holes exert an appreciable effect on the sound produced; the size or position of the lower open holes makes essentially no difference. It

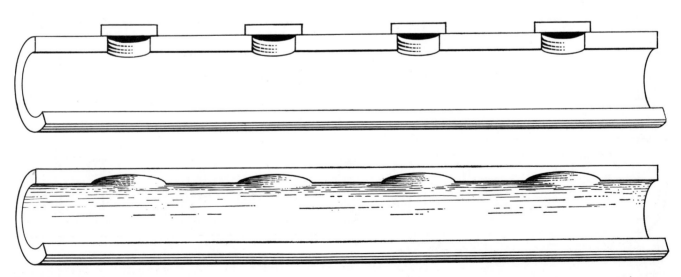

CLOSED HOLES alter the vibrational properties of the bore. Side holes closed by player's fingers or by pads on key levers   (*top*) **convert the bore from a smooth-walled pipe to a lumpy duct that may be thought of as a pipe with a series of swellings (*bottom*).**

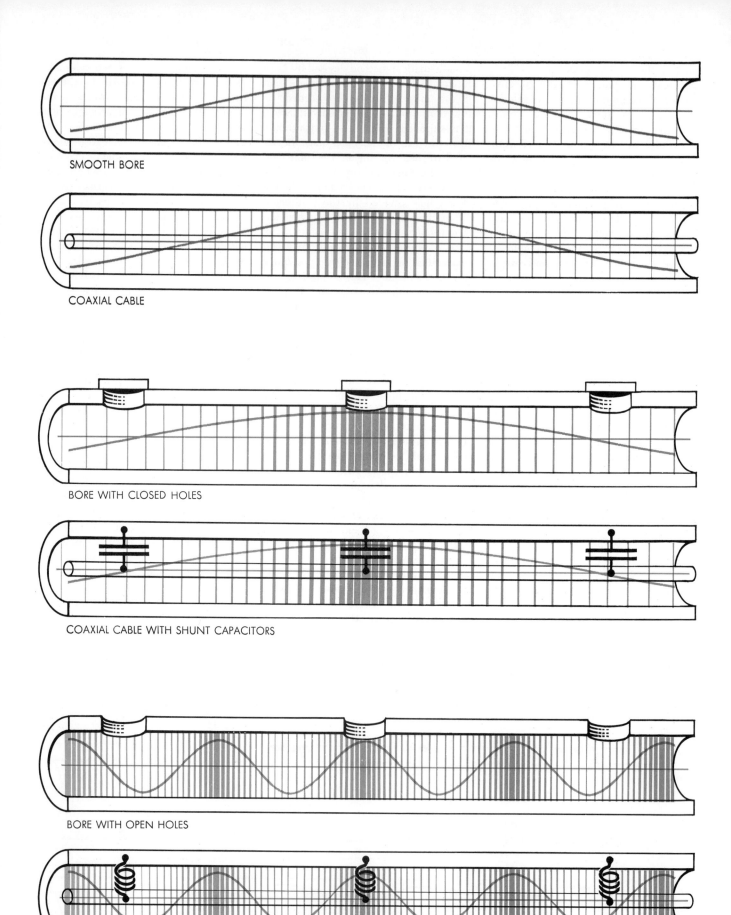

SMOOTH BORE

COAXIAL CABLE

BORE WITH CLOSED HOLES

COAXIAL CABLE WITH SHUNT CAPACITORS

BORE WITH OPEN HOLES

COAXIAL CABLE WITH SHUNT INDUCTORS

ELECTRICAL ANALOGY depicts a wood-wind bore as a coaxial cable. The acoustical vibrations of the air within a smooth-walled bore are analogous to the electrical vibrations in the cable at top.

Bore with closed side-holes behaves like cable with evenly spaced shunt capacitors connected across it (*middle*). Bore with open side-holes behaves like cable with evenly spaced inductors (*bottom*).

occurred to me that one might learn something by pretending that the bore contains open holes of uniform size and spacing, the dimensions being fixed by the size and spacing of the two highest holes that are open for the note being played. The relation between this simplifying abstraction and the pattern of holes in a real instrument is shown in the illustration on page 40 . To my delight, the trick succeeded in predicting the pitches of my own clarinet from measurements of its holes. After further experimental checks and a mathematical analysis as to why the whole informal scheme held together, I found that the main musical frequencies that give the pitch of a note are not able to travel very far in a bore with open holes; the larger, more widely spaced holes in the lower end of the bore send only weak "messages" back to the main bore.

William Dent has recently worked over this problem in a different way as a senior thesis project at Case Institute of Technology, using a mathematical approach that was originally suggested by Rayleigh, but was further developed for the purposes of quantum theory by Gregor Wentzel of the University of Chicago, H. A. Kramers of the University of Utrecht and Leon Brillouin. Dent found that in spite of the nonuniform arrangement of the open holes, they act almost as though they were of regular size and spacing, provided they are properly proportioned for their closed-hole duty. In short, although the lack of communication between the bore and the lower open holes makes the pitch relatively insensitive to their size and spacing, the messages that do get through from these holes make it seem as though they were uniformly spaced after all. Of course uniform spacing is not necessarily a musical virtue, but it is a convenience that makes it possible to use the mathematical methods that are at hand.

The relatively easy success of my first attempts encouraged me to re-examine the role of the closed holes in influencing the vibrations of the bore. It became clear that the closed holes act as a filter that discriminates strongly against the highest few components of the vibration spectrum produced by the reed. The "cut-off" frequency of this filter depends critically on the size and location of the closed holes. In flutes and saxophones the cut-off frequency is so high that it has little effect on the tone of the instrument, but in all the other wood winds (especially the oboe) the tone color is considerably altered by the filtering effect of the closed holes.

The next step was to try coaxing Mason's transmission-line equations into giving information about the way in which vibrations in the bore are coupled to the outside air. Any electrical or acoustical engineer looking at the equations for the row of open holes would instantly recognize them as describing a second sort of filter which transmits high-frequency vibrations, but attenuates those of lower frequencies. A wood wind thus emits the lower components of its tone into a room rather inefficiently, chiefly from the first one or two holes but symmetrically in all directions from the instrument. The higher components are radiated efficiently from all the open holes acting in concert, but in a highly directional manner. The size and spacing of the holes determine the frequency at which the "cross-over" in the mode of radiation takes place. In this complicated fashion the holes help determine the tone color of the sound that reaches the ears of the listener.

Listeners have come to associate wood winds with the type of sound that is emitted (by each of the two radiation mechanisms) from a row of open holes. The addition of the bell at the lower end of some wood winds reflects efforts over the years to provide the bore with a radiating system that approximates the behavior of a row of open holes even when all the holes are closed. It is a matter of long experience that this can never be done perfectly. (Of course an instrument maker might escape simultaneously from tradition and from the problem by simply providing a few extra open holes at the bottom of the bore which would be used only as emitters of sound.) In contrast to the clarinet and oboe, the flutes and saxophones are essentially bell-less because they radiate all components of their tones as if from a single hole.

This brief account of the interwoven complexities of wood-wind instruments has suggested some of the ways in which their structure affects their behavior. But it also suggests a more general observation. The curious weaving of familiar knowledge from various apparently unrelated fields, illuminated by flickers of intuition and analogy, is typical of the way in which most scientific knowledge develops. While many scientific efforts are shaped by esthetic considerations, the physics of music is particularly fortunate in being allied to an art from which it draws inspiration, and to which it often brings a deeper understanding.

5

The Physics
of Brasses

by Arthur H. Benade
July 1973

*A trumpet produces musical tones when the vibrations
of the player's lips interact with standing waves in the
instrument. These waves are generated when acoustic
energy is sent back by the instrument's bell*

It is easy to grasp why stringed instruments make the sounds they do. When the strings are struck or plucked, they vibrate at different natural frequencies in accordance with their tension and their diameter. The energy of vibration is then transferred to the air by way of a vibrating plate of wood and a resonating air chamber, with the sound eventually dying away. The musician can vary the pitch, or frequency, of individual strings by changing their vibrating length with the pressure of his fingers on the frets or the fingerboard.

The principles underlying the acoustics of bowed-string instruments such as the violin or wind instruments such as the oboe are a good deal less obvious. Here a vibration is maintained by a feedback mechanism that converts a steady motion of the bow, or a steady application of blowing pressure, into an oscillatory acoustical disturbance that we can hear. On the violin and in the oboe different tones are produced by altering the effective length of the string or the air column.

Like the oboe and other woodwinds, the brass instruments can produce sustained tones. The question arises, however, of how a bugle, which is hardly more than a loop of brass tubing with a mouthpiece at one end and a flaring bell at the other, can produce a dozen or more distinct notes. Horns were fashioned and played for centuries before physicists were able to work out good explanations of how they worked, even though scientific attention has been directed to these questions from the earliest days. For centuries the skilled craftsman has usually been able to identify what is wrong with faulty instruments and to fix them without recourse to sophisticated knowledge of horn acoustics.

All brass instruments consist of a mouthpiece (which has a cup and a tapered back bore), a mouthpipe (which also has a carefully controlled taper), a main bore (which is either cylindrical or conical) and a flaring bell that forms the exit from the interior of the horn into the space around the instrument. Brass instruments are of two main types. Those in one family, which includes the trumpet, the trombone and the French horn, have a considerable length of cylindrical tubing in the middle section and an abruptly flaring bell. Those in the other family, called conical, include the flügelhorn, the alto horn, the baritone horn and the tuba. The generic term conical refers to the fact that much of the tubing increases in diameter from the mouthpiece to the bell and the flare of the bell is itself less pronounced than it is in the first family. Actually all the horns called conical incorporate a certain amount of cylindrical tubing in their midsection. Here I shall deal primarily with the properties of instruments in the trumpet and trombone family. The properties of the conical instruments are very similar except for being somewhat simpler acoustically because overall they have much less flare.

The acoustical study of waves in an air column whose cross section varies along its length (a "horn") goes back to the middle of the 18th century. Daniel Bernoulli, Leonhard Euler and Joseph Louis Lagrange were the first to discuss the equations for waves in such horns during the decade following 1760. Their activity was a part of the immensely rapid blossoming of theoretical physics that took place in the years after the laws of motion had been formulated by Newton and Leibniz. Theoretical investigations of fluid dynamics, acoustics, heat flow and the mechanics of solid objects took their inspiration from the workaday world outside the laboratory and the mathematician's study. The work of Bernoulli, Euler and Lagrange on horns (and their similar researches on strings) did not have much influence in the long run on the science of acoustics or the art of music. It was nonetheless a part of the initial blooming of the theory of partial differential equations underlying nearly all physics.

The "horn equation," as we call it today, was neglected until 1838, when George Green rediscovered it while investigating the erosion caused by waves in the new canal systems of England. Then the equation was buried again until 1876, when a German mathematician, L. Pochhammer, independently derived it for waves in a column of air and learned the properties of its most important solutions. Neither Pochhammer nor his equation was long remembered. Finally in 1919 an American physicist, A. G. Webster, published a report on the horn equation, with the result that the equation is commonly named for him.

Since Webster's time interest in loudspeakers on the part of the phonograph and radio industries, to say nothing of military demands for sonar gear to detect submarines, has kept the subject of horn acoustics in a lively state. A loudspeaker horn must be designed to radiate sound efficiently out into the air over a broad range of frequencies from a small source. A horn designed to serve as a musical instrument has quite different requirements. In a musical horn the flare of the bell must be designed to trap energy inside the horn, giving strongly marked standing waves at precisely defined frequencies.

It is obvious that as a wave travels into the enlarging part of a horn its pressure will decrease systematically, simply because the sound energy is being spread over an ever wider front. If one extracts this intuitively obvious part of the behavior of a wave in a horn from the mathematics of the horn equation, one is left with a much simpler equation that is identical in form with the celebrated Schrödinger equation of quantum mechanics. The Schrödinger equation shows that a particle of energy E has associated with it a de Broglie wavelength lambda (λ) that depends on the square root of the difference between the energy and the potential energy function V at any point in space. The "reduced," or simplified, form of the horn equation shows similarly that at any point in the horn the acoustic wavelength depends on the square root of the difference between the squared frequency and a "horn function" U that depends in a rather simple way on the nature of the horn flare [see top illustration on next page].

It is not difficult to show from the horn equation that sounds propagate with dif-

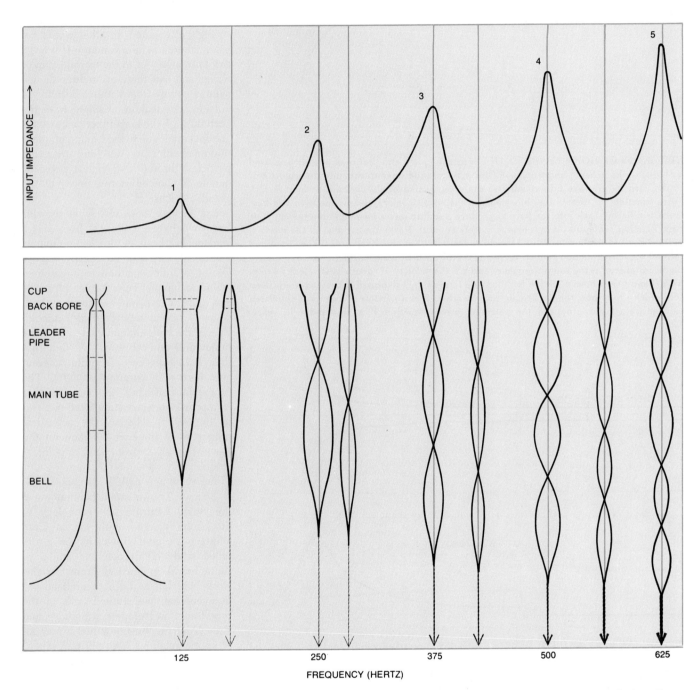

RESONANCE PEAKS OF A TRUMPETLIKE INSTRUMENT can be plotted (top) in terms of the impedance measured at the mouthpiece. Impedance is defined as the ratio of the pressure set up in the mouthpiece to the excitatory flow that gives rise to it. The impedance depends on whether the sound wave reflected from the bell of the horn returns in step or out of step with the oscillatory pressure wave produced in the mouthpiece. The shape of the air column in the trumpetlike instrument is shown at the extreme left of the bottom part of the diagram. The curves at the right are the standing-wave patterns that exist in the air column of the instrument at frequencies that produce the maxima and minima in the impedance curve. The first maximum is at about 100 hertz (cycles per second), when the reflected wave is precisely in step with the entering wave. The small irregularities in the standing-wave pattern are produced by the abrupt changes in the cross section of the instrument. The first minimum comes just above 125 hertz, where the returning wave and the incoming wave are exactly out of step with each other in the mouthpiece of the instrument. The subsequent maxima and minima are similarly explained. The number of nodes in the standing-wave pattern increases by one at each impedance peak.

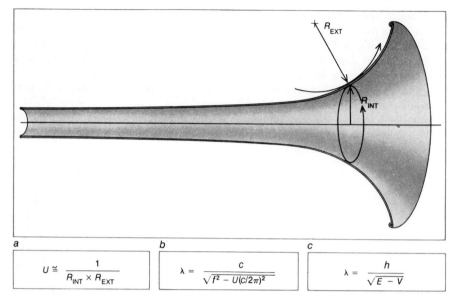

$$U \cong \frac{1}{R_{INT} \times R_{EXT}}$$

$$\lambda = \frac{c}{\sqrt{f^2 - U(c/2\pi)^2}}$$

$$\lambda = \frac{h}{\sqrt{E - V}}$$

GEOMETRY OF HORN FLARE largely governs the pitch and timbre of sounds produced by horns of the trumpet and trombone family. As a sound wave travels into the flaring bell of the horn its pressure falls steadily as the cross section of the instrument increases. A "horn function," U, determines how much of the acoustic energy leaves the horn and how much is reflected back into the horn to produce standing waves inside the instrument. The horn function (*equation "a"*) is approximately equal to 1 over the product of the internal radius (R_{int}) of the horn and the external radius (R_{ext}) at any given point. The simplified form of the horn equation (*equation "b"*) gives the acoustic wavelength (λ) at any point in the horn, where f is the sound frequency and c is the velocity of sound. This velocity varies with U and f. The horn equation has the same form as the celebrated Schrödinger equation (*c*), which shows how the de Broglie wavelength (λ) of a particle of energy E is related to Planck's constant (h) and the potential energy function V at any point in space.

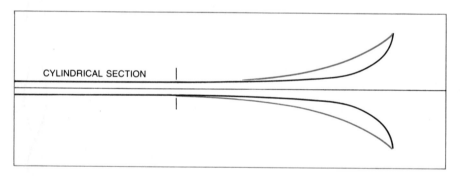

CYLINDRICAL SECTION

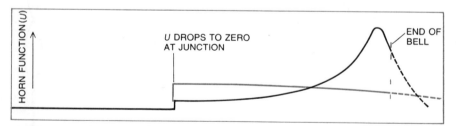

HORN FUNCTION (U)

U DROPS TO ZERO AT JUNCTION

END OF BELL

TROMBONE BELL AND LOUDSPEAKER HORN are markedly different in geometry and acoustic properties. The catenoidal shape (*black curve at top*) of the loudspeaker horn favors the efficient radiation of sound into the air. The flaring shape (*colored curve at top*) of the trombone bell is designed to save energy inside the horn, thus generating strongly marked standing waves at closely defined frequencies. Both the trombone bell and the loudspeaker horn are shown attached to a short section of cylindrical pipe. The two curves at the bottom show the horn function, U, for each horn. The catenoidal horn has a horn function (*colored curve*) that is low and nearly constant except for a slight falling off at the large end, where the sound wave fronts begin to bulge appreciably. The horn function (*black curve*) of the trombone bell rises steeply and falls. The higher the value of the function U, the higher the barrier to sounds of low frequency. Sounds of higher frequency are able to progress farther before they are reflected back by the barrier. In both cases above a certain frequency most of the sound energy radiates over the top of the barrier, so that the bell of the trombone loses its musically useful character and behaves like a loudspeaker horn.

ferent speeds as they travel through regions of differing horn function U. The speed of propagation also depends on the frequency. Another similarity between horn acoustics and quantum mechanics is that for frequencies below a certain critical value determined by the magnitude of U, the wavelength becomes mathematically imaginary, or, to put it in more physical terms, the wave changes its character and becomes strongly attenuated. In other words, regions where the horn function U is large can form a barrier to the transmission of waves and can therefore reduce the escape of energy from within a horn to the outside. The leaking of sound from the horn through the horn-function barrier is an exact analogue to the leaking of quantum-mechanical waves (and therefore particles) through the nuclear potential barrier in the radioactive decay of the atomic nucleus.

Let us look more closely at the difference between a musical horn and a loudspeaker horn. A simple example of a musical horn can be constructed by joining a trombone bell to a piece of cylindrical pipe. To a similar pipe one can join a typical loudspeaker bell, whose figure is described as catenoid. Even if the bells are matched to have the same radii at both ends, we find that their horn functions are quite different [*see bottom illustration at left*]. The catenoidal bell has a horn function that is approximately constant from one end to the other, whereas the acoustical properties of the horn function for the musical horn vary from point to point.

Five years ago Erik V. Jansson of the Speech Transmission Laboratory of the Royal Institute of Technology in Stockholm worked with me at Case Western Reserve University on a detailed study of air columns similar to those found in musical horns. In this work, which was both theoretical and experimental, we studied bells of the type found on trumpets, trombones and French horns. We unearthed a number of subtle relations between our experiments and calculations that we did not have time to clarify immediately. It is only recently that we have had an opportunity to prepare complete reports on our results. In what follows I shall lean heavily on information gained in our work five years ago and its later development, and on the earlier observations of many people concerned with acoustics or making musical horns.

In a brass musical instrument the small end of the horn is connected to the

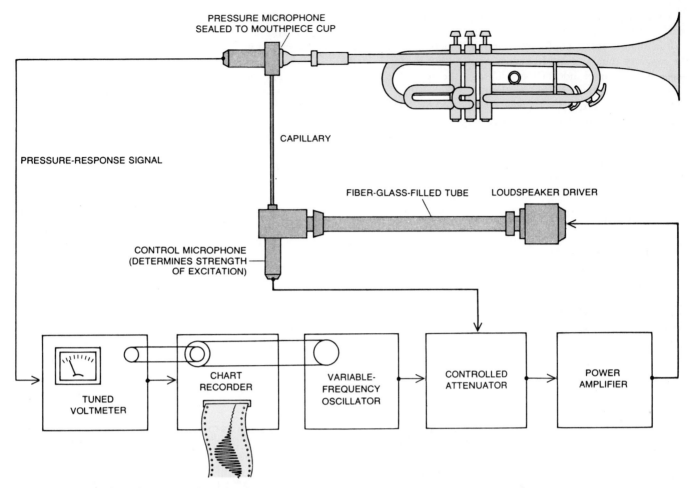

IMPEDANCE-MEASURING APPARATUS uses the driver from a horn loudspeaker as a pump to feed a flow stimulus through a capillary into the mouthpiece cup of the instrument under study. A control microphone sends signals to an attenuator to ensure that the acoustic stimulus entering the capillary remains constant. The pressure response of the instrument, and thus its input impedance, is detected by a second microphone that forms the closure of the mouthpiece cup. The signal from the microphone goes to a frequency-selective voltmeter coupled by a chain drive to oscillator. A chart recorder coupled to the voltmeter plots the resonance curves.

player through his lips, which constitute a kind of automatically controlled valve for admitting air from the player's lungs to the horn. The opening and closing of the valve is controlled chiefly by the pressure fluctuations within the mouthpiece as they act on the lips in concert with the steady pressure from the lungs. Therefore an initial objective is to find the relations between the flow of air into the horn and the acoustical pressure set up at the input end.

Let us begin by imagining a laboratory experiment in which the horn is excited not by air from the player's lips and lungs but rather by a small oscillatory flow of air being pumped in and out of the mouthpiece through a fine capillary by a high-speed pump. This small oscillatory flow disturbance in the mouthpiece gives rise to a pressure wave that ultimately reaches the flaring part of the horn. As the wave travels down the length of the bore of the horn some of its energy is dissipated by friction and the transfer of heat to the walls of the

instrument. In the flaring part of the bell a substantial fraction of the acoustic wave is reflected back toward the mouthpiece while the remainder penetrates the horn-function barrier and is radiated out into the surrounding space. The wave that is reflected back down the bore of the horn combines with newly injected waves to produce a standing wave.

If the round-trip time that the wave takes to go from the mouthpiece to the bell and back to the mouthpiece is equal to half the repetition time of the original stimulus or to any odd multiple of the repetition time, a standing wave of considerable pressure can build up and result in a large disturbance in the mouthpiece. At intermediate frequencies of excitation the return wave tends to cancel the influence of the injected wave. In other words, depending on the precise interaction between the injected wave and the reflected wave, the pressure disturbance inside the mouthpiece can be large or small. For purposes of describing such disturbances in the mouthpiece

under conditions of constant flow excitation in a laboratory apparatus, engineers define a quantity termed input impedance: the ratio between the pressure amplitude set up in the mouthpiece and the excitatory flow that gives rise to it [see illustration on page 45].

The shape of the horn controls the natural frequencies associated with the various impedance maxima and minima by determining the penetration of the standing waves into the bell. The shape also controls the amount of wave energy that leaks out of the horn into the surrounding space. Furthermore, the kinks in the standing wave that arise from discontinuities in cross section and taper along an air column produce significant changes in both the resonance and the radiation properties of the bell. The interaction of the kinks and the primary shape of the air column can spell the difference between success and failure in the design of an instrument.

There are several ways one might measure the input impedance, or response, of the air column. Conceptually the simplest method would be to pump air in and out of the mouthpiece through a capillary tube at some frequency and measure the amplitude of the resulting pressure fluctuations in the mouthpiece by means of a probe microphone. It is more practical, however, to use the driver of a commercial horn loudspeaker as a pump. The motion of the driver is controlled electronically by an auxiliary monitor microphone that maintains a constant strength of oscillatory flow through the capillary as one sweeps automatically through the appropriate range of frequencies. Between 1945 and 1965 Earle L. Kent and his co-workers at C. G. Conn Ltd. in Elkhart, Ind., developed this basic technique to a high degree of dependability. We often employ a modification of their technique in our work [see illustration on preceding page].

In Cleveland we make use of two ad-ditional methods that have special advantages for certain purposes. The first method, based on a device described in 1968 by Josef Merhaut of Prague, can be applied in measurements not only on the smaller brasses but also on bassoons and clarinets [see illustration below]. In Merhaut's device a thin diaphragm forms a closure at the end of the mouthpiece cup and itself serves as the pump piston. The diaphragm is driven acoustically through a pipe that connects it to an enclosed loudspeaker. The diaphragm motion is monitored for automatic control by the electrode of a condenser microphone mounted directly behind it. The second method is based on a device that was used by John W. Coltman of the Westinghouse Research Laboratories in investigating the sounding mechanism of the flute. In Coltman's device the excitatory diaphragm is driven directly by a loudspeaker coil whose motion is monitored by means of a second pickup coil

that is moving in an auxiliary magnetic field [see illustration on opposite page].

If one attaches to any one of these excitation systems a cylindrical section about 140 centimeters long from a trumpet, one discovers dozens of input impedance peaks evenly spaced at odd multiples of about 63 hertz (cycles per second) [see curve "a" in top illustration on page 50]. The peaks correspond exactly to what elementary physics textbooks describe as the "natural frequencies of a cylindrical pipe stopped at one end." Because frictional and thermal losses inside the tube walls increase with frequency, the resonance peaks become smaller at higher frequencies. The energy radiated from the open end of such a pipe is only a tiny fraction of 1 percent of the wall losses.

If one now adds a trumpet bell to the same cylindrical pipe, the impedance response curve is substantially altered [see curve "b" in bottom illustration on

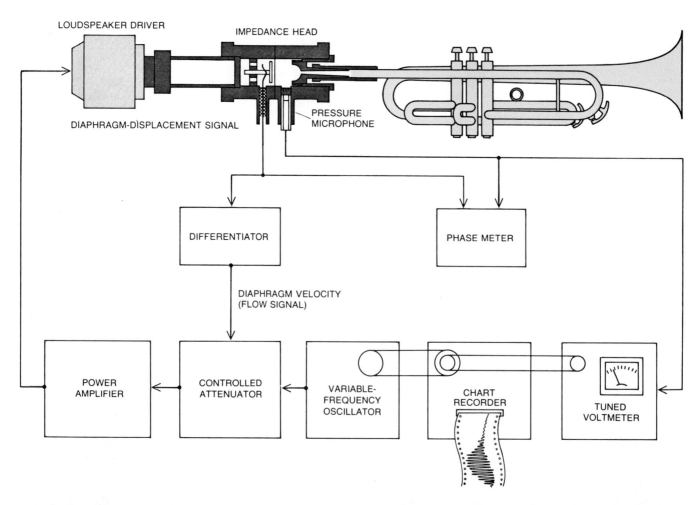

SECOND TYPE OF IMPEDANCE-MEASURING DEVICE was developed by Josef Merhaut. It differs from the apparatus illustrated on the preceding page only in the way that the flow stimulus into the mouthpiece is controlled. Here the acoustic stimulus produced by a loudspeaker moves an aluminized Mylar diaphragm that in turn pumps air into the mouthpiece. The diaphragm also acts as one electrode of a condenser microphone to produce a signal proportional to the diaphragm's velocity and thus proportional to the oscillatory flow of air at the mouthpiece cup. The velocity signal adjusts the attenuator in order to maintain constant excitation at a particular frequency. The pressure response of the instrument is monitored by a microphone on the cup side of the diaphragm. A phase meter shows the relation between the phase of the input stimulus and the phase of the pressure response of the instrument.

next page]. The first peak is hardly shifted at all by adding the bell, but the frequencies of the other resonances are lowered in a smooth progression because the injected waves penetrate ever more deeply into the bell before being reflected. In addition the peaks at higher frequencies are markedly reduced in height because a growing fraction of the energy supply leaks through the bell "barrier" as the frequency is increased. In sum, the return wave in the pipe-plus-bell system is weakened not only by wall losses but also by radiation losses, particularly at high frequencies. Above about 1,500 hertz essentially no energy returns from the flaring part of the bell. The small wiggles in the impedance curve at high frequencies are due chiefly to small reflections produced at the discontinuity where the bell joins the cylindrical tubing.

By comparing these curves for incomplete instruments with the impedance curve for a complete cornet [*see illustration on page 51*] one can see at a glance that the presence of a mouthpipe and mouthpiece has a considerable effect on the overall nature of the input impedance. The resonance peaks of the cornet grow taller up to around 800 hertz, then fall away much more abruptly than the curve produced by the pipe-plus-bell system.

Let us now consider how the player's lips control the flow of air from his lungs into the instrument. As the player blows harder and harder, the flow increases both because of the increased pressure across the aperture formed by his lips and because his lips are forced farther apart by the rising pressure inside his mouth. Equally important is the variation imposed on the flow by pressure variations inside the mouthpiece, which tend to increase or decrease the flow by their own ability to affect the size of the lip aperture. It is this pressure-operated flow control by the lips under the influence of the mouthpiece pressure that ultimately leads to the possibility of self-sustained oscillation. Let us abstract from this rather complicated situation only the relevant part of it: the alteration in net flow that is produced by acoustical pressure variations within the cup of the mouthpiece. As long ago as the middle of the 19th century it was clearly understood that it is the flow alteration due to mouthpiece pressure that can maintain an oscillation.

In 1830 Wilhelm Weber described experiments on the action of organ reeds

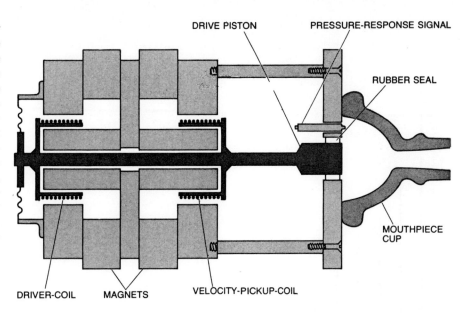

ELECTROMAGNETIC SOURCE for projecting acoustic waves into a test instrument was devised by John W. Coltman. The excitatory piston is directly coupled to the voice coil of a loudspeaker. The coil in turn drives the piston with an amplitude that is ultimately determined by a voltage induced in a pickup coil that is mounted on the same shaft. The mechanism is used in an overall system similar to that used with the Merhaut impedance head. The pressure response in the mouthpiece cup is detected by a miniature microphone.

that led him to a correct theory for the effect of a compliant structure (the reed or, in our case, the player's lips) on the input impedance of a column of air. This effect of the yielding closure of the mouthpiece cup provided by the lips is quite separate from the lips' functioning as a valve. Hermann von Helmholtz provided the next advance. In 1877 he added an appendix to the fourth German edition of his classic work *Sensations of Tone* that gives a brief but complete analysis of the basic mechanisms by which a pressure-controlled reed valve collaborates with a single impedance maximum. He found that for a given pressure-control sensitivity (what an engineer today calls the transconductance) a certain minimum impedance value is required. Oscillating systems of the type analyzed by Helmholtz are found around us everywhere. The pendulum clock is possibly the oldest and most familiar. The wristwatch, electronic or otherwise, falls into this category. Every radio and television set has one such oscillator or more.

Engineers have studied oscillating systems intensively and have learned that even if the alteration in flow (of whatever kind) that results from a given pressure is not exactly proportional to the pressure (as Helmholtz assumed for simplicity in his pioneering investigation) but varies in some more arbitrary fashion, the properties of the system are not drastically altered. The presence of such

nonlinearity in the control characteristics gives rise to additional frequencies at double, triple and quadruple the frequency of the basic oscillation. The net generation of oscillatory energy from the player's steady muscular effort, however, is still almost exclusively at the frequency of the impedance maximum; energy diverted in the process to other frequencies is dissipated in various ways to the outside world.

We must now try to explain how oscillations in a wind instrument can take place at not just the tallest impedance maximum but at any one of several maxima belonging to an actual air column. According to the Helmholtz theory, a wind instrument should show a strong preference for oscillations that take place at the tallest of the impedance maxima. Thus the question arises of how the bugle player finds it possible to play the notes based on lesser impedance maxima. Furthermore, one must ask how the bugler is able to select one or another of these peaks in accordance with his musical requirements.

It is not in fact difficult to deal with the problem of how the player selects one note or another. His lips are so massive compared with the mass of the air in his instrument that the influence of the air column on the lips is relatively small. The player adjusts the tension of his lips in such a way that their own natural tendency of vibration favors oscillation at the desired note, so that the

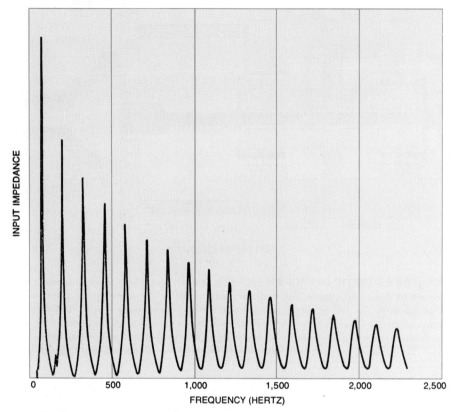

IMPEDANCE PATTERN OF SIMPLE CYLINDRICAL PIPE 140 centimeters long shows peaks evenly spaced at odd multiples of 63 hertz. The higher the frequency, the greater the loss of wave energy to the walls of the pipe through friction, hence the steady decline in the height of the peaks. Less than 1 percent of the input energy is radiated into the room.

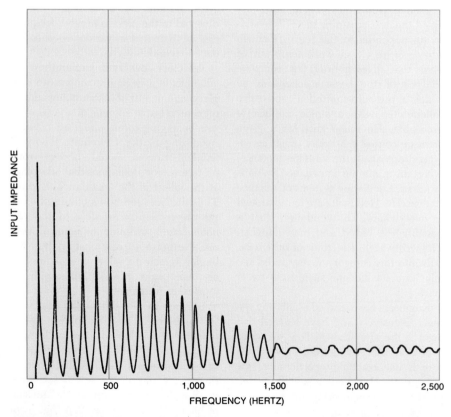

ADDITION OF TRUMPET BELL TO PIPE lowers the overall height of the impedance peaks and squeezes them together. Whereas the pipe alone produces 16 peaks in a span of 2,000 hertz, the pipe-plus-bell system compresses the first 16 peaks into a span of 1,400 hertz. Beyond 1,500 hertz more and more of the acoustic energy leaks through the bell barrier.

air column and the lips collaborate in producing the desired frequency.

So far we have not said anything that could not have been understood in terms of 19th-century acoustics. The best account of the Weber-Helmholtz analysis and its musical consequences was made by a French physicist, Henri Bouasse, in his book *Instruments à Vent*, the two volumes of which appeared in 1929 and 1930. These volumes contain what still constitutes one of the most thorough accounts of the acoustics of wind instruments, encompassing the flute and reed organ pipes, the orchestral woodwinds and the brasses. Bouasse has left us with a gold mine of mathematical analysis, along with an account of careful experiments done by himself in collaboration with M. Fouché or selected from the writings of earlier investigators.

Bouasse was quite aware of the inadequacy of a theory of oscillation assuming that all the energy production is at the basic frequency of oscillation. He described many phenomena observed among the reed organ pipes and the orchestral woodwinds and brasses that underscore the limitations of this general viewpoint and that imply cooperation among several air-column resonances. Bouasse's interest in these matters was to serve both as a strong incentive and as an invaluable guide when I later undertook a close study of the subject. The first fruits of this study were described in a series of technical reports written in 1958 for C. G. Conn Ltd.

By 1964 I found it possible to deal well enough with the interaction between a reed valve and an air column having several impedance maxima that I could design and build a nonplaying "tacet horn." This "instrument" has several input impedance maxima chosen in such a way as to make them unable to maintain any oscillation in cooperation with a reed, even though the Weber-Helmholtz theory would predict the possibility of oscillation. In 1968 Daniel Gans and I published an account of this theory of cooperative oscillations. That report, based on Gans's undergraduate thesis at Case Western Reserve, included a description of the tacet horn and explanations of various phenomena discussed by Bouasse. Since that time the work has been carried much further in our laboratory, particularly by Walter Worman, who wrote his doctoral dissertation on the theory of self-sustained oscillations of this multiple type in 1971. Although his work was focused on clarinetlike systems, his results apply broadly to all the wind instruments, including the brasses. These studies were aided by counsel from many people, in particular Bruce Schantz, Kent, Robert W. Pyle, Jr., and John H. Schelleng.

It is now time to see how the Weber-Helmholtz form of the theory had to be modified, using the trumpet as our example. When the musician sounds one of the tones of a trumpet, the air column and his lips are functioning in what we shall formally call a regime of oscillation: a state of oscillation in which several impedance maxima of the air column collaborate with the lip-valve mechanism to generate energy in a steady oscillation containing several harmonically related frequency components. Worman was able to trace out how a set of impedance maxima can work together with the air valve. The particular "playing frequency" chosen by the oscillation (along with its necessarily whole-number multiples) is one that maximizes the total generation of acoustic energy, which is then shared among the various frequency components in a well-defined way.

Experiments with instruments as diverse as the clarinet, the oboe, the bassoon, the trumpet and the French horn show that softly played notes are dominated by the impedance maximum that belongs to the note in the sense of Weber and Helmholtz. As the musician raises the dynamic level, however, the influence of the higher resonances grows in a definite way that is common to all the instruments. As he plays louder and louder, the influence of the impedance at double the playing frequency becomes more marked, and for still louder playing the resonance properties at triple or quadruple frequencies join the regime of oscillation one by one. A look at the input impedance curves for a modern trumpet will show how the peaks in a regime of oscillation cooperate so that the player can sound various notes on his instrument, including even some notes that have no peak at all at the playing frequency [*see illustrations on next two pages*]. Notes in this last category have been known to brass players since the earliest days and were a part of horn-playing technique in the time of Mozart and Beethoven. The need for such notes was reduced, however, as the instrument became more mechanized. In recent years they have returned; for example, they are sounded by musicians who want to play bass-trombone parts without resorting to a special thumb-operated valve that is otherwise required. Tuba players also find the technique useful on occasion.

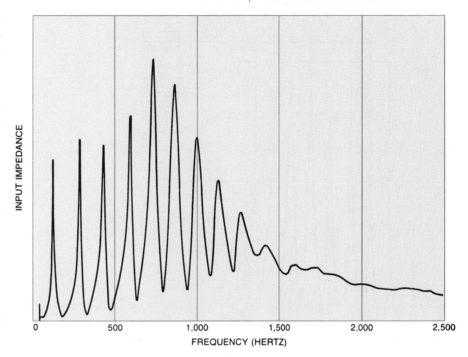

IMPEDANCE PATTERN OF A 19TH-CENTURY CORNET is typical of most of the trumpet and trombone family. The peaks grow progressively and then fall away sharply. The cornet was made in 1865 by Henry Distin. The third and fourth impedance peaks do not quite follow the smoothly rising pattern required for a genuinely fine instrument. The shortcoming is due chiefly to slight constrictions and misalignments in the valve pistons.

The reader may be wondering what happens when the valves on a brass instrument are depressed. Does anything radically new happen? The answer is no. The bell, the mouthpipe and the mouthpiece dominate the "envelope," or overall pattern, of the resonance curve; the pattern of peaks for a trumpet rises steadily as one goes from low frequencies to about 850 hertz and then falls away and disappears at high frequencies. When a valve is depressed, thereby increasing the length of cylindrical tubing in the middle of the horn, it merely shifts the entire family of resonance peaks to lower frequencies but leaves them fitting pretty much the same envelope.

In addition to working out the details of the regimes of oscillation in wind instruments Worman gained an important insight into the factors that influence tone color. He was able to show that in instruments with a pressure-controlled air valve (a reed or the lips) the strength of the various harmonics generated in a regime of oscillation (as measured inside the mouthpiece) has a particularly simple relation when the instrument is being played at low and medium levels of loudness. Let us take as given the strength of the fundamental component that coincides with the playing frequency. As one would expect, that strength increases as the player blows harder.

Worman's striking result is that when the player blows very softly, there is essentially no other component present in the vibration as it is measured in the mouthpiece, and that as he plays louder the amplitude of the second harmonic grows in such a way that for every doubling of the strength of the fundamental as the player blows harder, the strength of the second harmonic quadruples. Furthermore, the strength of this component proves to be approximately proportional to the impedance of the air column at the frequency of the second harmonic. Similarly, the third harmonic has a strength that is proportional to the impedance at the third-harmonic frequency, and from an even tinier beginning it grows eightfold for every doubling of the strength of the fundamental component. In short, the nth harmonic has a strength that is proportional to the impedance at the nth harmonic of the playing note, and that component grows as the nth power of the fundamental pressure amplitude. The remarkable thing about Worman's observation is that it is totally independent of all details of the flow-control properties of the reed or the lips, provided only that the flow is controlled solely by the pressure variations in the mouthpiece [*see top illustration on page 54*].

Let me summarize what we have found out so far about how the tone

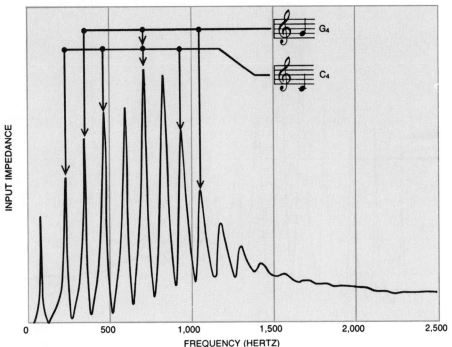

IMPEDANCE PATTERN OF A MODERN TRUMPET is annotated to show what happens when a player sounds the notes C_4 or G_4. When he blows into the horn, a "regime of oscillation" is set up in which several impedance maxima of the air column collaborate with oscillations of his lips to generate energy in a steady oscillation that contains several harmonically related frequency components. The regime of oscillation for the C_4 note involves the second, fourth, sixth and eighth peaks in the curve. When the trumpeter plays very softly, the second peak is dominant, but because this peak is not tall the beginner may produce a wobbly note. As he plays louder the other peaks become more influential and the oscillation becomes stabilized. The dominant oscillation for the G_4 note corresponds to the third impedance peak; since it is taller than the second peak, G_4 is easier than C_4 to play pianissimo. As the trumpeter plays louder the tall sixth peak comes in and greatly stabilizes the regime of oscillation, making the G_4 one of the easiest notes of all to play.

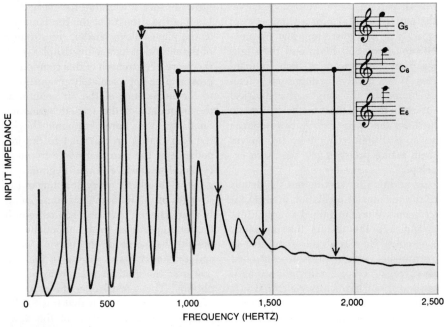

REGIMES OF OSCILLATION FOR HIGHER NOTES show why they become increasingly hard to play as one moves up the scale. G_5 is still quite easy to play because its regime of oscillation is dominated by the tall sixth impedance peak; the 12th peak makes only a minor contribution. C_6 is somewhat more difficult to play because the dominant peak of the note is lower than the peak for G_5. It takes an athletic trumpeter to reach the high E_6 and higher notes. The trumpet at this point has become virtually a megaphone: the energy production of the instrument is due almost completely to the interaction of the air column with the lips themselves, much as the human larynx operates in producing vocal sounds.

quality develops as measured inside the mouthpiece of the brass instruments. When one plays very softly, only the fundamental component associated with the playing frequency is present. As one plays louder the second, third, fourth and still higher harmonics grow progressively. If the oscillation is in the nature of a regime involving several cooperating resonance peaks, the harmonics grow in the simple fashion described by Worman's theorem; it is only at very loud playing levels that his theorem fails to give simple results. Furthermore, the theorem shows that the strength of the various components is proportional to the height of the various impedance maxima that are cooperating to generate the tone. In other words, when one plays rather loud, the strengths of the various harmonics have heights that correspond roughly to the heights of the impedance maxima from which they draw their chief sustenance. On the other hand, when a tone is generated on the basis of only a single resonance peak, as is the case in the upper part of the trumpet's range, we would be able to describe the strength of the components only if we could specify all the details of the flow-control characteristic.

Up to this point I have been discussing only the strength of the various harmonics as they are measured by a small probe microphone inside the brass instrument's mouthpiece cup. What one hears in the concert hall is, of course, very different. The transformation from the spectrum generated inside the mouthpiece, where the actual dynamics of the oscillation are taking place, into the spectrum found in the concert hall has to do with the transmission of sound from the mouthpiece into the main air column and thence out through the bell. There are many facets to the total transmission process, even without taking into account the complexities of room acoustics or the complications of our perceptual mechanism, which does a remarkable job of processing the great irregularity of room properties to give us clearcut, definite impressions of the tone quality of musical instruments. I shall only remark that the transformation of the spectrum inside the mouthpiece to the external spectrum has the general nature of a treble boost. In other words, whatever sounds may be generated inside the instrument, it is the higher components that are radiated into the room [see *bottom illustration on page 54*].

The very fact that the bell of an instrument leaks energy preferentially at high frequencies has two important consequences. On the one hand the leakage enhances the relative amount of high-frequency energy that comes out of the horn; on the other it serves to reduce the height of the impedance peaks at high frequencies that lead to the weak generation of the high-frequency part of the spectrum inside the instrument. As a result measurements made outside the instrument in a room do not show nearly as much instructive detail about the dynamics of the entire system as measurements made inside the instrument do.

Let me conclude this discussion of the physics of brass instruments by indicating some of its implications for the musician and the instrument maker. As an illustration of the way physics can help the musician, I shall quote from an article of mine that appeared recently in the magazine *Selmer Bandwagon*. In this passage it was my intention to help French-horn players clarify and systematize their technique of placing one hand in the bell of the instrument to enhance certain frequencies.

"The player's hand in the bell is, acoustically speaking, a part of the bell. ... A properly placed hand provides... resonance peaks out to 1,500 hertz on an instrument that otherwise would lose all visible peaks at about 750 hertz [*see illustration on page 55*]. Suppose you meet a totally unfamiliar horn (perhaps during a museum visit when the curator opens the display cases) and you wish to find out quickly how well the instrument plays. Blow a mid-range note (for example concert F_3 in the bass clef) and, keeping your hand absolutely flat and straight, push it into the bell little by little until you feel a slight tingle in your fingertips. At this point (keeping the hand always perfectly straight) move the hand in and out a little until the horn sings as clearly as possible and the oscillation feels secure to your lips. Any listening bystander will agree with your final choice. Keep your hand in this slightly strained position and blow a tone an octave or a twelfth above the first one (say concert F_4 or C_5). Keeping your fingertips always in their original position, bend the palm of your hand so that its heel moves toward a position more familiar to the horn player. As you do so the tone will again fill out and get a ringing quality to it; also your lips will vibrate with a more solid feel. Your hand will now be in an excellent position for playing all notes on this horn, although an expert will be able to do even better after careful practice.

"Moving your straightened hand in and out while sounding the low *F* allows you to arrange to have an accurately located second helper for the tone. The unstopped horn works somewhat like a trumpet playing G_5 above the staff, while putting in the flattened hand serves to set up a regime that is analogous to the one which runs the trumpet's midstaff C_5. Bending the palm of one's hand while keeping the fingertips in place will leave the resonance peaks adjusted so far pretty much intact, but will make them taller (and hence more influential). This also gives rise to more peaks at the high-frequency end of things. The frequencies of these peaks move as the hand is bent more, so that once again the player has a means for tuning them for optimum cooperation with the other members of the regime. Trumpet players sometimes find it interesting and technically worthwhile to adapt the horn player's hand technique for their own purposes—especially for playing high passages on a piccolo trumpet."

It is only in the past few years that we have begun to have an understanding of the acoustics of mouthpieces. William Cardwell of Whittier, Calif., has provided a good theoretical basis for dealing with the relation of the mouthpiece dimensions to the tuning of the various resonance peaks. We in Cleveland, with the help of George McCracken of the King Musical Instrument Division of the Seeburg Corporation, have given attention to how the mouthpiece design controls the height of the impedance peaks. I quote again from the article for musicians to indicate the practical implications of mouthpiece acoustics.

"Acoustical theory tells us that, first and foremost, a given instrument will require that the mouthpiece have a certain well-defined 'popping frequency' when its cup is slapped shut against the palm of the hand. In other words, the lowest natural frequency of the mouthpiece alone (with the cup closed) must be of the correct value. It is this requirement

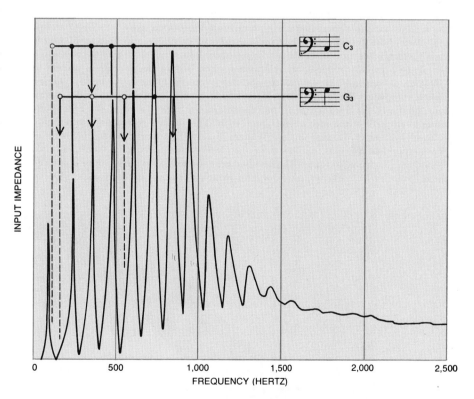

UNUSUAL REGIMES OF OSCILLATION are associated with notes whose frequencies correspond to impedances that are close to minimum values. The note C_3 in the bass clef is known to musicians as the pedal tone. Its regime of oscillation is such that the second, third and fourth resonance peaks of the trumpet sustain an oscillation that lies at a frequency equal to the common difference between their own natural frequencies. Since there is actually a loss of energy at the fundamental playing frequency for this note rather than a gain, there is only a small amount of fundamental component in the sound, and even the small quantity present is converted to that frequency from the higher components by way of the nonlinearity in the flow-control characteristics of the player's lips. The situation for G_3 is even more unusual in that the second and fourth components of the tone are the chief source of oscillatory energy, whereas the fundamental component and the other odd harmonics contribute virtually nothing since the impedance is minimal at their frequencies.

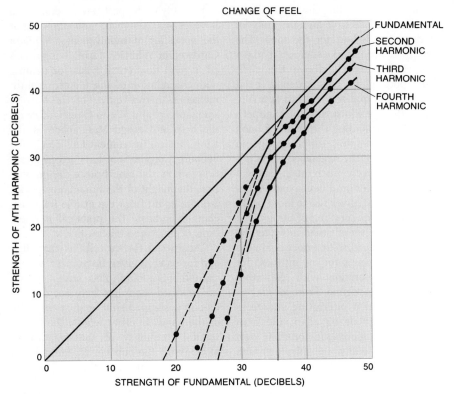

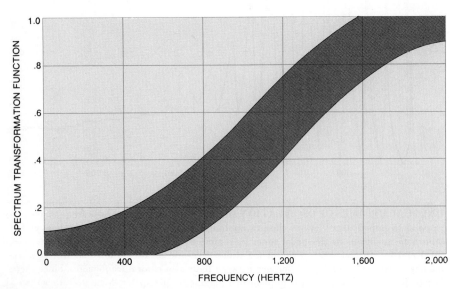

TONE COLOR OF TRUMPET is related to the way harmonic frequencies make up an increasing fraction of the total sound emitted as the player blows louder. The strengths of the various harmonic components are plotted as a logarithmic scale (decibels) against the logarithm of the strength of the fundamental component. At low and medium playing levels each harmonic lies on a straight line whose slope is approximately equal to the serial number of the harmonic. As one plays pianissimo essentially no harmonics are present in the vibration as measured in the mouthpiece. For every doubling in strength of the fundamental component the second harmonic increases from an initial tiny value by a factor of four. Similarly, the third harmonic increases in strength by a factor of eight for each doubling in strength of the fundamental, and so on. This finding corresponds to a theory developed by Walter Worman at Case Western Reserve University. At the loudness where Worman's relation begins to break down the player senses a change in "feel" and listeners are aware of a change in sound. The data that are reflected in the curves were obtained with the help of Charles Schlueter, who now plays principal trumpet in the Minnesota Orchestra.

TRANSMISSION OF TRUMPET SOUND INTO ROOM is characterized by the "spectrum transformation function," which indicates what fraction of the acoustic energy at each frequency, as measured inside the mouthpiece, is emitted from the bell. Depending on the level of play and characteristics of the instrument, the energy emitted usually falls within the band plotted here. The curve has the qualitative nature of a "treble boost" because the bell leaks energy preferentially at high frequencies. Numbers on vertical scale are arbitrary.

that determines which of the peaks in the trumpet's response curve are the tallest. It also helps the peaks in this region to have the proper frequencies for good cooperation with the low-note regimes. The second most stringent requirement on the mouthpiece is that its total volume be correct (cup plus backbore). We must have this volume right in order to make the bottom two or three regimes of oscillation work properly."

So far I have discussed only the factors that contribute to favorable oscillation inside the horn and have said nothing about the tuning of instruments in the musician's sense: the relation between the pitches of the various tones that the instrument will generate. Fortunately the requirements for good tuning are almost identical with the requirements for favorable oscillation. It is for this reason that the traditional musical-instrument maker, focusing the major part of his attention on the tuning of the notes of the instrument, was able to develop instruments that would "speak" well and have good tone.

In more recent years, as our knowledge of acoustics has grown and the computer has become available, efforts have been made to design good brass instruments with the computer's help. Here the influence of loudspeaker acoustics has been great. Substantial efforts have been made to mathematically piece together a sequence of short loudspeaker-horn segments, each one intended locally to represent the shape of a workable brass instrument. This segmental approach to the problem has certain computational advantages. As we have seen, wherever the bore of a horn has a discontinuity of angle or of cross section there are anomalies in the standing-wave pattern. In spite of this fact it is always possible in principle to find suitable angles and cross sections that will place the impedance maxima of the horn with an accuracy that is acceptable by tuning standards. Although instruments built in this manner may play fairly well in tune, they can be quite disappointing in their musical value because of the neglect of the more subtle cooperative phenomena that ultimately distinguish between mediocrity and genuine excellence. Furthermore, the ability of an instrument to speak promptly and cleanly at the beginning of a tone is extremely sensitive to the presence of discontinuities, so that even though these discontinuities are arranged to offset one another in such a way as to give an excellent steady tone, it does not follow that the instrument starts well. The musician must of course have a "clean attack" as well as a clear, steady tone.

The skillful instrument maker gradually acquires an almost instinctive feel for the subtleties of instruments, so that he can sometimes be astonishingly quick in the use of his empirical store of knowledge to find a correct solution to a tuning or response problem. Consider the problem that such a person must solve when he is asked to correct a trumpet that is faulty, with the sole error being the behavior of the tone corresponding to C_4. Let us suppose that the problem is caused by the fourth impedance peak (beginning from the peak of lowest frequency), which is somewhat high in its frequency. When the C_4 is played at a pianissimo level, the note will be in tune, but as the loudness increases somewhat the note will tend to run a little sharp as the second member of the regime (the mistuned fourth peak) begins to show its influence. The player will also notice that he can "lip" the tone up and down over a considerable range in pitch without appreciable change in tone color. He will complain that at this moderate dynamic level the tone "lacks center." If he plays louder, the influence of the still properly tuned third and fourth members of the regime becomes strong enough to partly overcome the defect of the second member. When this occurs, the player finds that the tone once again acquires what he calls a core, or center, at a certain playing level, which happens then to fall pretty well back in tune because all but one of the resonances in the regime agree on the desired playing pitch.

In the practical world of the instrument maker or designer one often meets instruments in which one or more notes are "bad" in this way. It has often proved quite difficult to correct such problems with only instinct and experience. Once one understands what is going on, however, it is often possible to bypass laboratory measurements and diagnose the errors with the help of carefully designed "player's experiments." One then uses acoustical perturbation theory to guide the alteration of the shape of the air column to give a desired correction. Such corrections are made by enlarging or reducing the cross section of the bore in one region or more of the air column. The problem is complicated by the need to preserve the locations of the correctly tuned resonance peaks while the faulty peak is being moved.

Whether one is a physicist, a musician or an instrument maker, one tries to make use of any tools at hand to provide an instrument that helps rather than hinders the creative effort of music making. At first it would seem that the computer is ideally suited to be one of these tools and that it could immediately be put to work designing the perfect instrument. As a practical matter one finds that although we have a reasonable understanding of the goals to be achieved, the complexity of the problem is such that it is very difficult to specify the problem for the computer in sufficient detail. I have found that it is much more efficient to start with an already existing good instrument developed by traditional methods and then apply the physical understanding and the technical facilities available to us today to guide the improvement of the instrument, whether it is for an individual player in a symphony orchestra or for the development of a prototype for large-scale production.

In all my work I have found it always important to keep in constant touch both with professional players and with instrument makers. They provide an inexhaustible supply of information about the properties of instruments. They also are a source of questions that have proved enormously fruitful in guiding my investigations. As the subject continues to develop it is becoming increasingly possible for the results of formal acoustical research to be translated into useful information for the player and the instrument maker.

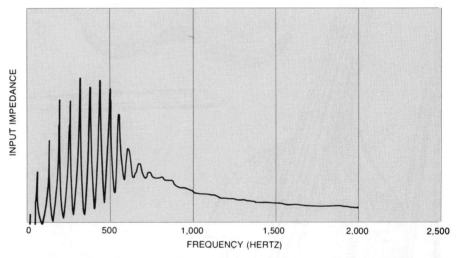

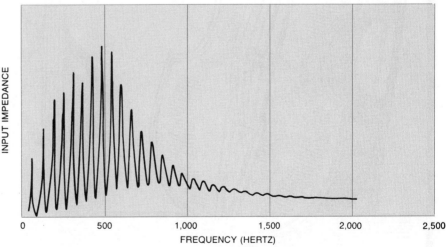

PLACING HAND IN BELL OF FRENCH HORN is a well-known technique for extending the frequency range of the instrument. The curve at the top shows the input impedance response of a valveless prototype for the *B*-flat half of a standard French horn when measured without the player's hand in the bell. There are essentially no resonance peaks above 750 hertz. If the player tries to reach a note such as G_5 (783 hertz), all he gets is a wobbly scream because there is little or no feedback of acoustic energy from the bell of the instrument to stabilize a note of higher frequency. Notes in the octave below G_5 would also be weak and characterless for lack of a strong feedback. The curve at the bottom shows the additional resonance peaks produced when the musician points his flattened hand into the bell until he feels a slight tingling at his fingertips and then bends his palm slightly. The instrument now produces peaks well beyond a frequency of 1,000 hertz, making it possible for the musician to play the note G_5 quite dependably and even a few higher notes when he is pressed.

VIOL FAMILY, a somewhat earlier development than the violin family that eventually supplanted it, encompassed a large number of instruments of varying shapes and sizes. This drawing from the *Syntagmatis Musici* of Michael Praetorius (1619) shows three examples of the viola da gamba (*1, 2, 3*), a viola bastarda (*4*) and a viola da braccio (*5*). They have not all been drawn to the same scale.

The Physics
of Violins

by Carleen Maley Hutchins
November 1962

*Modern acoustics is making it possible to account for
the exquisite performance of the violins made by the
Italian masters. The results promise a further evolution
in the instruments of the violin family*

During the Renaissance there grew up in Italy two new families of musical instruments, both of them stemming from primitive stringed instruments of the Middle Ages such as the rebec and lute. The earlier of the two groups to emerge was the viols; the later, following about a century afterward, was the violins. The violin was not an outgrowth of the viol but a somewhat later development from similar sources, and the two were lively competitors for a long time. Composers wrote distinctive music for each kind of instrument, and each had its virtuoso performers. Eventually the violin family, having a richer and more powerful sound, supplanted the older group—except for the largest and lowest-pitched instrument, which survives as the bass viol.

As this story unfolds it will be clear that viols still have more than mere historical interest. For the moment I shall describe the viols briefly. They came in a variety of shapes and sizes [*see illustration on opposite page*], most of them having a flat back unlike the beautifully arched back plate of the violins. They had five, six or more strings, more slackly tuned than violin strings and supported on a flatter bridge. Often their finger boards were crossed by gut frets resembling the metal ridges on the finger board of a guitar. Their wooden sounding boards were lighter and more flexible than those of the violin family.

Exactly who invented the violin is not clear. It may have been Andrea Amati, who in any case founded the great Cremona school of violinmakers. Amati died around 1580; within 150 years or so his descendants and their pupils, particularly Antonio Stradivari and Giuseppe Guarneri, had brought the art of violinmaking to such an extraordinarily high level that it is only now that one dares

to dream of equaling or surpassing it. These early masters must have had an open mind toward the little that was known in their time about the physics of sound. Their successors deserve credit for having lovingly preserved an art, but certainly not for advancing a science. In effect they have formed a cult that has been plagued with more peculiar notions and pseudo science than even medicine.

Today the well-developed science of acoustics is applicable to the understanding and making of violins. For the past 30 years or so a handful of interested physicists, chemists, musicians and some people who, like me, began as amateurs have been applying it. In fact, we have organized ourselves informally as the Catgut Acoustical Society. Much of what

has been learned is still empirical, but it is nonetheless interesting and valuable. In this article I shall try to touch on at least the high spots of our studies.

In essence a violin—as well as its larger, deeper-voiced relatives, the viola and the cello (properly the violoncello) —is a set of strings mounted on a wooden box containing an almost closed air space. Some energy from the vibrations induced by drawing a bow across the strings (precious little energy, it turns out) is communicated to the box and the air space, in which are set up corresponding vibrations. These in turn set the air between the instrument and the listener into vibration; in other words, they produce the sound waves that reach his ears. That is the main story. The sound of a violin, putting aside the acoustics of

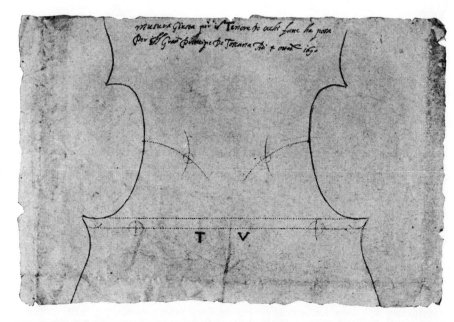

DRAWING BY ANTONIO STRADIVARI marks positions of the upper and lower ends of "f-holes" in a tenor viola. At top he has written: "Exact measurements for the sound holes of the tenor made expressly for the Grand Prince of Tuscany, the 4th day of October, 1690."

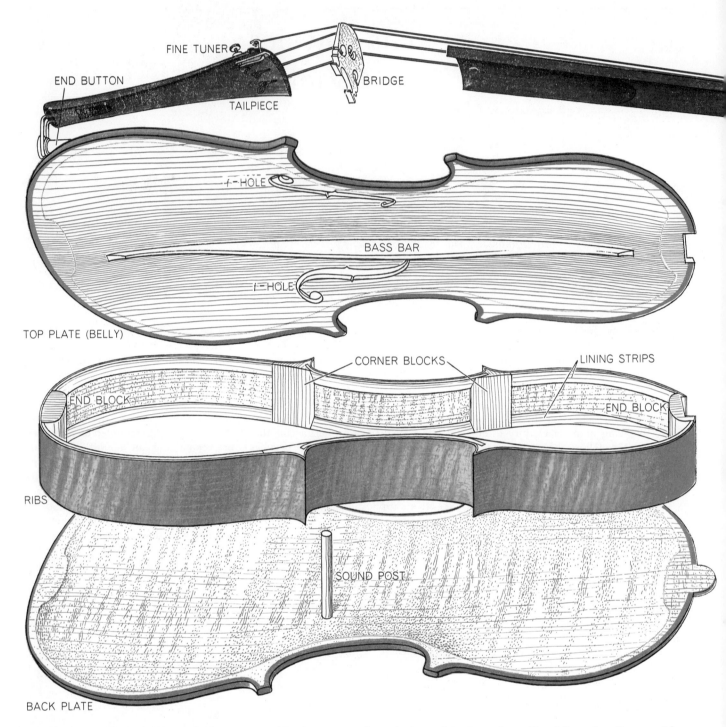

FINE TUNER

END BUTTON

TAILPIECE

BRIDGE

f-HOLE

BASS BAR

f-HOLE

TOP PLATE (BELLY)

CORNER BLOCKS

LINING STRIPS

END BLOCK

END BLOCK

RIBS

SOUND POST

BACK PLATE

ANATOMY OF VIOLIN INSTRUMENTS is essentially the same for the violin, viola and cello. The exploded view of the viola on these two pages shows the top plate, ribs, back plate and devices for stringing at left, the neck, scroll and finger board at top right. Immediately below, a section of the top plate illustrates the bilateral symmetry of the grain of the spruce wood. At bottom

the room in which it is played and the skill of the player, depends on the transfer of vibration from string to sounding box to air.

The Basic Violin

Before getting into this apparently innocent problem, which turns out to be a veritable jungle of unknowns, it is worthwhile to pause for a moment to examine the instrument itself. Violin strings are usually made of metal, pig gut or gut wound with fine silver or alumi-

num wire. The sounding box consists of a front plate and a back plate, both arched slightly outward to form broad bell-like shapes, and the supporting ribs, or sides. The back plate is carved with chisel, plane and scraper, traditionally from a block of curly maple seasoned for at least 10 years and not kiln-dried. (Pear or sycamore wood are sometimes used.) It can be a single piece or two pieces carefully joined. In thickness the back plate varies from about six millimeters in the center to almost two millimeters just inside the edges (from

1/4 inch to 5/64 inch). The sides are pieces of matching curly maple, thinned down to a millimeter all over, bent into shape and glued to spruce or willow blocks set in the corners and at the forward and rear ends of the plates.

The top plate, usually spruce, is split lengthwise from a log and then joined so that the wood of the outside of the tree is in the center of the top, making the grain bilaterally symmetrical. In thickness the top plate ranges from two to three millimeters, and a pair of beautifully shaped "f-holes" are cut into each

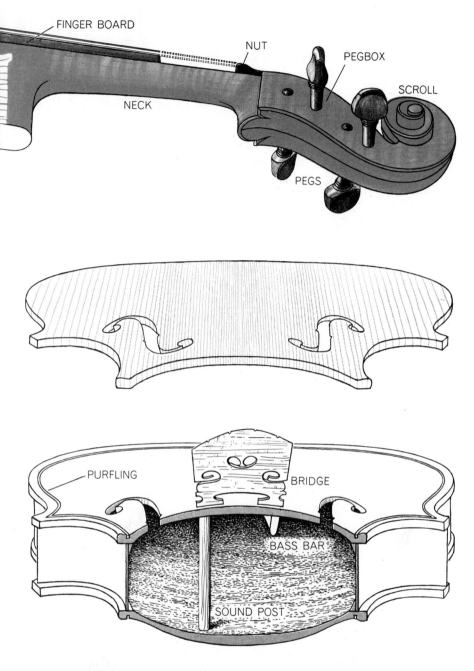

right a cross section through the middle of the instrument shows the relative positions of the bridge, bass bar and sound post. The purfling consists of three very narrow strips of wood that are set in a shallow groove around the edge of both the top and back plates.

open question. In a few years we hope to have some answers.

I might anticipate a bit here to mention a point that illustrates the subtlety of some of the problems in understanding the violin. Does the purfling serve any purpose other than decoration? It happens that the wood of the plates underneath the purfling is extremely thin. After years of playing, the glue that holds the purfling strips in their grooves begins to crack, in effect creating a vibrating plate with very thin edges. Frederick A. Saunders, professor emeritus of physics at Harvard University, has suggested that this may be a factor in the improved tone of an instrument that has been played for a long time.

The combined tension of the four strings of a properly tuned violin comes to around 50 pounds. As a result about 20 pounds is directed straight down through the bridge and against the delicate eggshell-like sounding box. To distribute the load and help the top plate withstand the downward component of string tension the viol makers glued to it a strip of wood running lengthwise down the middle. Whether by accident or by a stroke of genius, one of the earlier violinmakers moved the bar to one side so that one foot of the bridge rested above it. The strip, made of spruce, is now called the bass bar, since it is under the foot of the bridge on the side of the string of lowest tuning.

To support the other foot of the bridge there is placed approximately underneath it a vertical post, also made of spruce, called the sound post. It is carefully fitted and held in place between the front and back plates by friction. The acoustical function of the sound post has been a matter of debate for many years. The tone of a violin can be so greatly altered by small changes in the position, tightness and wood quality of the sound post that the French call it the soul (*l'âme*) of the instrument. Removing it altogether makes the violin sound rather like a guitar.

Although the modes of vibration of the plates exhibit great diversity throughout the frequency range, the bridge must always have some rocking motion to receive power from the string. In the important lower half of the range the sound post and the adjacent foot of the bridge have relatively little motion, thus providing in a sense a fulcrum that serves to transfer maximum travel to the bridge foot standing over the bass bar.

This too is getting a little ahead of the story, which begins when a bow is drawn across one or more of the strings of a violin. Vibrating strings have been stud-

side of the plate. All around the outside of each plate, near the edge, is cut a shallow groove in which is inlaid the "purfling," consisting usually of two strips of black-dyed pearwood and a strip of white poplar.

Other materials may be mentioned: curly maple for the neck, ebony for the finger board, rosewood or ebony for the tuning pegs and tailpiece, hard maple for the bridge. The outside of the instrument is treated with filler and varnished. Filler, varnish and glue all contribute to the over-all characteristics of the violin,

but there is no definite evidence to show that 300 years ago any of them was superior to the materials available now. In fact, the Catgut Acoustical Society is working to discover new substances that may be even more effective than the old. But it is a slow, painstaking search.

These were the general specifications for the Cremona fiddles and, with minor variations, for all good instruments since. Whether there is a mysteriously unique virtue in any of the woods or finishes, or whether some other types might not do as well for various purposes, is an

32.703 41.203 55.000 65.406 73.416 97.999 130.813 146.832 195.998 220.000 261.626 293.665

C₁ E A C₂ D G C₃ D G A C₄ D

EADG

ADGC

CGDA

LARGE
BASS

SMALL
BASS

NEW
CELLO

NEW VIOLIN FAMILY will consist of eight instruments, twice the number of the present violin family. Of these the large bass and the treble violin are yet to be constructed; the small bass and the tenor and soprano violins are newly designed instruments; the new cello and vertical viola are rescaled instruments; the violin is the only member of the original family to be included in the new

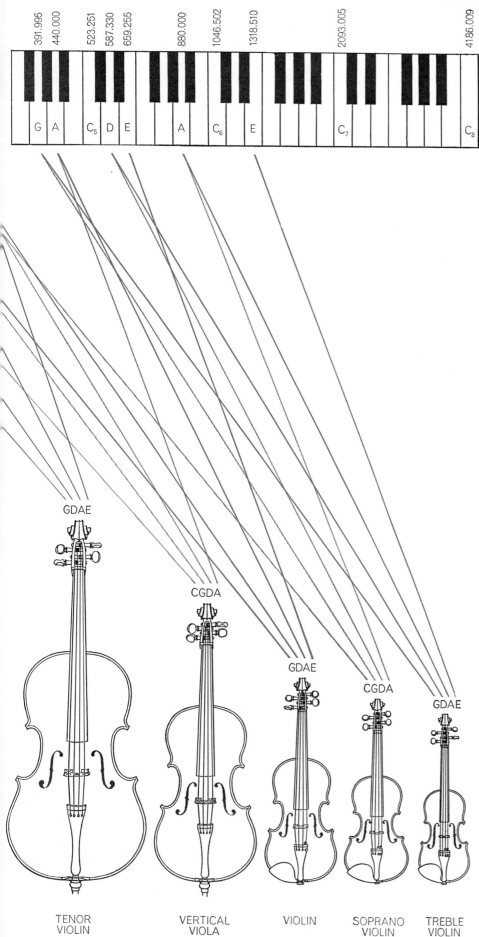

391.995	440.000	523.251	587.330	659.255	880.000	1046.502	1318.510	2093.005	4186.009

G A C_5 D E A C_6 E C_7 C_8

GDAE — TENOR VIOLIN

CGDA — VERTICAL VIOLA

GDAE — VIOLIN

CGDA — SOPRANO VIOLIN

GDAE — TREBLE VIOLIN

group unchanged (*see illustration on next page*). The notes to which the four strings of each instrument are tuned can be read from the piano keyboard and the associated colored lines indicating the strings. The numbers at top show the frequencies in cycles per second.

ied since the time of Pythagoras. An early 19th-century French physicist, Félix Savart, showed that the bowed string has a multitude of harmonics; then the great German physicist Hermann von Helmholtz elucidated the types of vibration that distinguish the bowed string from the plucked string. In our own century the Indian physicist Sir C. V. Raman made an exhaustive investigation of the vibration of bowed violin and cello strings. It would be fair to say that the reaction of a string to a bow is quite thoroughly understood.

In spite of the vigorous vibration of the moving string, the sound from the string alone would be all but inaudible. It has too little surface area to set an appreciable amount of air in motion. Trying to make music with an unamplified string would be like trying to fan oneself with a toothpick. What happens is that some portion of the energy supplied by the player to the bow—perhaps 5 to 10 per cent—is communicated to the wooden body of the violin through the complex motions of the bridge. (Of all the energy that the player feeds into the violin, 1 or 2 per cent emerges as sound. The rest goes off as heat.) The vibrations of the bowed string at any instant include dozens of energetic harmonics with amplitudes falling off as frequency increases. Each of the frequencies present shakes the wooden box—"forces" it to vibrate—at its particular rate. Obviously the amplitude of vibration depends on the strength, or amplitude, of the forcing vibration.

The Resonant Box

If this were all there was to it, matters would be simple, and all tones would be amplified equally. But the wooden structure itself has scores of frequencies at which it tends to vibrate naturally. The coincidence of such a frequency of resonance in the wood with the frequency of a string harmonic will result in an enhanced transfer of energy from string to box and a correspondingly greater amplification of that particular tone. Therefore the actual response of a violin to the playing of various notes is an enormously complex affair, but a good violinist must unconsciously and automatically deal with it and compensate for it every time he plays.

The scientific violinmaker is interested in all of these wood resonances, but he usually finds the resonance of lowest frequency an adequate guide during construction. This is called the main wood resonance. He is also interested in the lowest natural frequency (here only one seems to have any measurable importance) of the enclosed air space, called the main air resonance. Tests show that a good violin usually has its main wood resonance within a whole note of 440 cycles per second: the note A, to which the second highest string of the instrument is tuned.

Some instruments have a "wolf note," almost always at the frequency of the main wood resonance. When this note is played on any string, the tone warbles unsteadily, often breaking by a whole octave somewhat as the voice of an adolescent boy does. The wolf note occurs when the string and the wood form a pair of mechanically coupled circuits; a beating action occurs because energy is cyclically shuttled back and forth between them. Violas and cellos are notoriously subject to wolf-note trouble. Even some of the finest cellos have bad wolf notes. It is possible to ameliorate this difficulty in a variety of ways, for example by tuning a length of string between the bridge and the tailpiece to the actual frequency of the wolf note. This absorbs enough energy to control the wolf. So far, however, the ideal method of control has not been found.

Inside the box of the violin is the air chamber, or resonating cavity, which communicates directly to the outside by means of the f-holes in the top of the instrument. As I have said, the enclosed air has so far been found to add measurable resonance to a range of tones surrounding one note on each instrument. The pitch of this main air resonance, or air tone, can be approximately located by blowing across the f-hole, as one might blow across the top of an empty bottle. (When one f-hole is covered lightly, this pitch is lowered.)

The frequency of the air tone is controlled by the volume of air enclosed by the box of the instrument and the combined area of its f-hole openings. The larger the air volume, the lower the frequency; the larger the f-hole area, the higher the frequency. These two variables can be calculated roughly. I have found that to raise the resonance of the enclosed air a whole tone requires approximately a 20 per cent reduction in air volume or a 59 per cent increase in f-hole area. Anyone looking at the handsomely shaped f-holes of a violin can appreciate that it is not practical to try to raise the frequency of the air

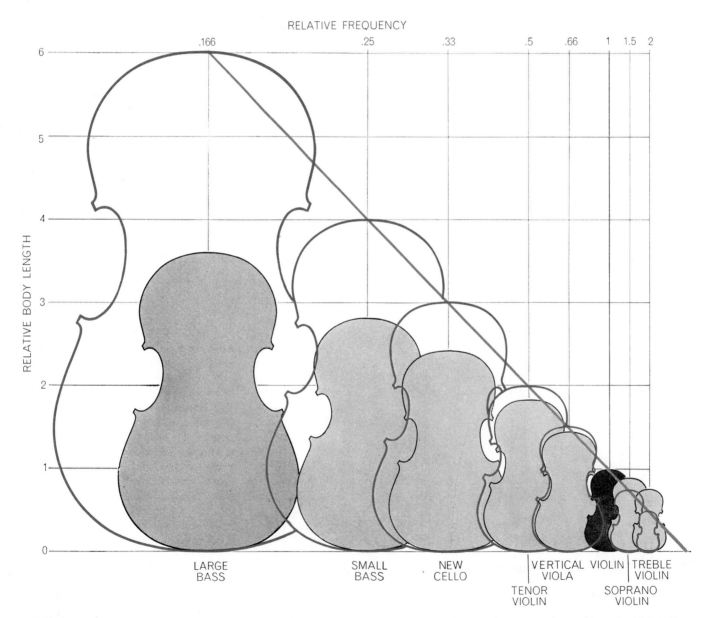

THEORETICAL AND ACTUAL SIZES of new instruments are compared. According to theory an instrument of a given relative frequency would have to be of the size represented by the colored outlines. In practice the size is that represented by the light gray areas. The relative frequency of an instrument (e.g., treble violin) is obtained by dividing the frequency of one of its strings (e.g., A) by the frequency of the corresponding string on the violin: 880 divided by 440 equals 2. The conventional viola, cello and double bass (*not shown in illustration*) have relative body lengths (using the violin as unity) of 1.17, 2.13 and 3.09 respectively.

resonance even a semi one by changing the size of the holes.

A number of workers—particularly Saunders, the late Hermann Backhaus of the Technische Hochschule in Karlsruhe, Hermann Meinel of Berlin, Gioacchino Pasqualini of Rome and E. Rohloff of the University of Greifswald—have developed methods of studying the resonances of violins. One of the most useful is the "loudness curve" originated by Saunders. It is also called the curve of total intensity, because it shows at each measured frequency the combined strengths of all the harmonics.

Loudness Curves

In making a loudness curve the violin is bowed normally, but without vibrato, at semitone intervals over its entire range to produce the loudest tone possible at each note. A General Radio sound-level meter, such as is used to measure levels of applause on television shows, records the loudness of each tone. It often comes as a shock to a musician to discover that his instrument is much louder at certain notes than at others. Try as he will he cannot possibly make them all register at an equally high level on the sound-level meter.

At the right are displayed loudness curves of a few violins, good and poor. In the good one the main wood resonance and the main air resonance fall approximately seven semitones, or a musical fifth, apart. (A fifth is the interval from "do" to "sol" on the diatonic scale. The frequency of each note is in the ratio of three to two for the note below it.) In some poor instruments the main wood and air resonances may be as much as 12 semitones, or an octave, apart (frequency ratio, two to one), giving two areas of strong resonance with a wide range of weak response between. The curve for one poor instrument shows only one area of strong resonance—the air resonance—with the wood contributing virtually nothing in the way of resonant reinforcement. A $5 violin with a curve almost as bad was used by Saunders for some time as his "standard" of badness. When I took the wretched thing apart and balanced it for good tone production, it showed an overall increase of loudness and an even spacing of peaks. At this point it was named Pygmalion. When it was played behind a screen in alternation with an excellent Cremona violin, the two were voted equal in tone by a college music department audience. In fairness it should be added that the skilled musician playing behind the screen was never in any doubt

as to which was the superior instrument.

An octave below the main wood resonance there is almost always another strong peak of loudness that we label "wood prime." It can be called a subharmonic. It is well known in acoustics that if one harmonic of a complex tone

is strengthened, the ear will hear an increase in loudness of the note as a whole with a slight change in quality but no change in pitch. By this process the wood peak is strengthened by the tone of the main wood resonance an octave above. The subharmonic of the main wood reso-

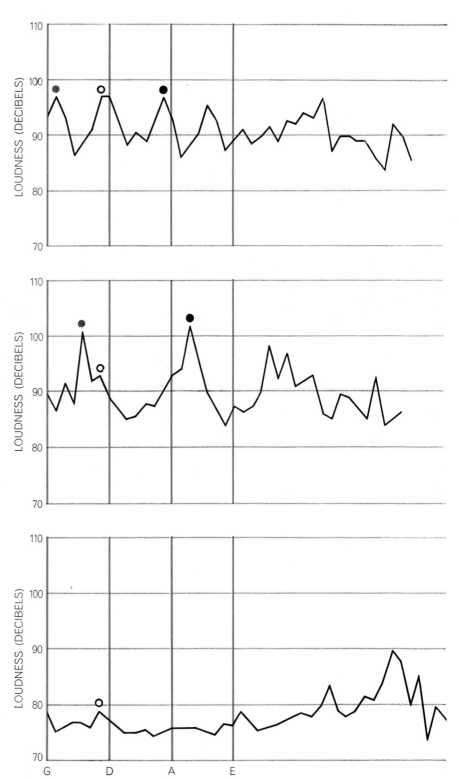

VIOLIN "LOUDNESS CURVES" compare maximum sound levels produced at semitone intervals by a good 1713 Stradivarius (*top*), a poor 250-year-old violin of doubtful origin (*middle*) and a poorer, somewhat older instrument credited to P. Guarneri (*bottom*). Only the first shows desirable spacing and strength of wood (*black dot*), "wood prime" (*gray dot*) and air resonances (*open circle*). Letters at bottom indicate tuning of open strings.

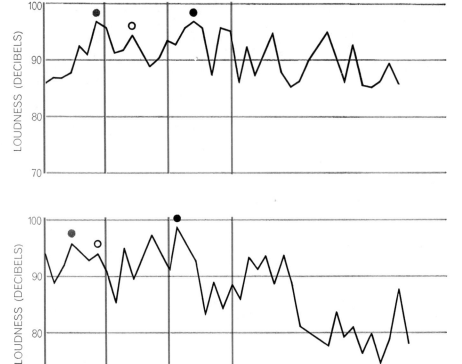

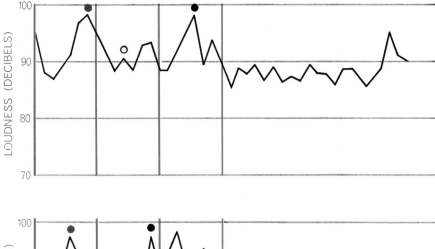

VIOLA LOUDNESS CURVES compare the responses of a conventional (*top*) and a vertical viola (*bottom*), both made by the author. The convention of dots and colored lines used in this illustration and that below is the same as in the illustration on preceding page.

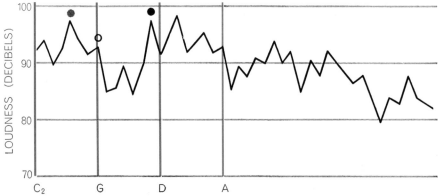

CELLO LOUDNESS CURVES compare the responses of a conventional (*top*) and a new cello (*bottom*), both made by the author. Note that the main wood and air resonances of the new viola and cello are near the two open middle strings. The loudness curves on this and preceding page are based on tests performed by Frederick A. Saunders of Harvard University.

nance benefits the lower tones of the violin, viola and cello.

The subharmonic of the main air resonance does not show on curves of conventional instruments because it falls below the bottom notes of the instruments. Spacing the main wood and air resonances about a half-octave apart spreads these peaks so that the air-tone peak falls nicely in the middle of the octave between the wood resonance and its subharmonic. In hundreds of tests of violins, violas and cellos this arrangement of wood and air resonances emerges as one of the characteristics of the good instruments.

Experimental Instruments

I have built a series of experimental violins and violas to test the effect of moving the frequencies of the main resonances up or down the scale. In a pair of violas of similar pattern with identical f-holes, one was made with sides half an inch high to decrease the air volume; the other had sides two inches high, giving a large air volume. Normally the sides of a viola are about 1½ inches high, and the air tone is found to be in the range from B to B flat (233 cycles per second) on the G string. In the viola with the smaller air volume the air tone, as expected, moved up the scale to D sharp (300 cycles per second). In the one with the larger air volume the air tone moved downscale near A (220 cycles per second).

In both of these altered violas the normally strong tones of the B to B flat on the G string were missing, because the air resonance was no longer there to reinforce them. Musicians playing the instruments discovered interesting features. Neither one was suitable for playing the two-viola quintets composed by Mozart. The composer had written so well for the normally strong tones of the viola that the oustanding parts lacked their full expressive qualities when the experimental instruments were used. The strong resonance of the air tone was not where musicians expected to find strength, nor where Mozart had counted on it.

The most interesting feature of the two violas was that the thin, shallow instrument had a full, rich tone and a particularly strong, low C string, where the normal viola is notably weak. This was because the air tone had been shifted upscale enough so that its subharmonic came into useful range near 150 cycles per second on the low C string.

The thick viola with the two-inch ribs,

on the other hand, had a thin tone, and the lower range of its C string was weak, partly because the air tone had been moved from its normal position. Many musicians playing the two violas in alternation have remarked with astonishment at the full, rich tone of the thin one with the small air volume. Ribs half an inch high are structurally not very practical, but application of the principles involved has made possible the construction of good small violas.

In studying the resonances of violins I have discovered that in the best violins the main wood and air resonances invariably fall within a semitone or two of the frequency of the two open middle strings, the wood resonance corresponding to the higher-tuned string. When the early violinmakers hit on this arrangement, the muses must have been smiling. It is, quite simply, the way in which most good violins have been made ever since.

This is not true of the viola and cello. In these instruments as they are now built the wood and air resonances fall three to four semitones higher with respect to the frequencies of the open middle strings than they do in the violin. The reason is simple enough: the viola and cello are built smaller than optimum size to make them a convenient playing size. As a result the resonances are too far above the lower notes of the instruments, and these suffer in strength and quality. I shall have more to say about this matter later.

Tap Tones

At the moment I should like to consider a different problem. Assuming one knows what the violin should be like when it is finished—where its resonances should fall and so on—how are these aims achieved in the process of construction? How does one make it come out the way one wants it? In addition to careful workmanship and accurate measurements the traditional method of the violinmaker has been to listen to the "tap tones" of the front and back plates.

In the final thinning and graduating of the top and back plates of a violin, the maker traditionally holds the plate near one end in his thumb and forefinger, taps it at various points with a knuckle and listens carefully to determine the pitch of the sounds he hears. These sounds are called the tap tones of the plates. The ability to judge the proper relation of tap tones of the free top and back plates in this manner is an important part of the art of violinmaking. With the ear alone it is extremely difficult to make out the frequencies of these tones, particularly in the case of the top plate, where the complicated structure with f-holes and bass bar creates at least two, and sometimes as many as five, strong natural resonances below 600 cycles per second.

Saunders and I, together with Alvin Hopping of Lake Hopatcong, N.J., have developed a method that makes it possible to determine the tap-tone frequencies in a free plate with considerable

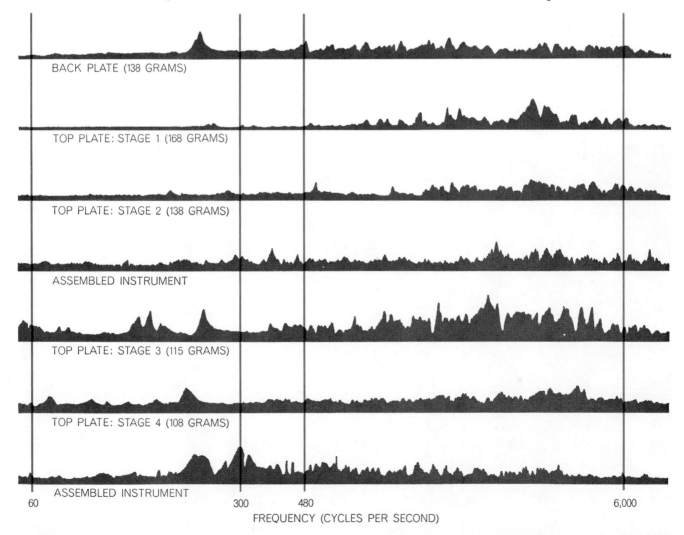

BACK PLATE (138 GRAMS)

TOP PLATE: STAGE 1 (168 GRAMS)

TOP PLATE: STAGE 2 (138 GRAMS)

ASSEMBLED INSTRUMENT

TOP PLATE: STAGE 3 (115 GRAMS)

TOP PLATE: STAGE 4 (108 GRAMS)

ASSEMBLED INSTRUMENT

60   300   480   6,000

FREQUENCY (CYCLES PER SECOND)

FREQUENCY-RESPONSE CURVES of top and back plates of a viola and of the assembled instrument at various stages are depicted. Although the tests run from 20 to 20,000 cycles per second, most of the response to the magnetic driver used to vibrate the wood falls in the range of 60 to roughly 10,000 cycles per second. The four frequencies indicated here are those at which checks were made to ensure that the recording film was synchronized with the audio-generator. Height of peaks represents amplitude of response.

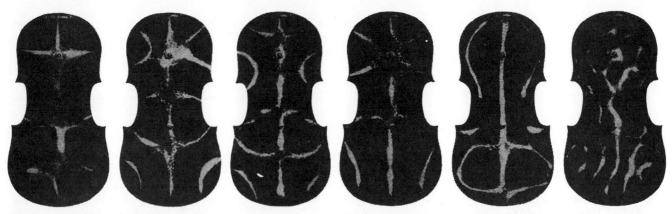

CHLADNI PATTERNS discussed in the text were made on a brass plate by Saunders at frequencies of 260, 340, 435, 520, 780 and 1,600 cycles per second. The plate is supported horizontally by a bolt at the center of its upper half; bottom end rests on a padded block.

accuracy. First we cut a flat brass plate in the shape of the violin plate. We dust it with powder and bow it at various points around the edge to set up different modes of vibration. Where the plate vibrates, the powder is bounced away, piling up along the nodal lines where there is no vibration. From these "Chladni" patterns on the brass [see illustration above] we are able to predict where a principal nodal point in the frequency test range will fall on the mid-line of a real violin plate. Since clamping on a nodal point does not affect the vibration pattern, we then clamp the violin plate at this point and set it into vibration at its exact center

by means of a magnetic driver, activated by an audio-frequency generator, that can be varied from 20 to 20,000 cycles per second. The response of the wood plate to the input signal, which has variable frequency but constant amplitude, is picked up by a microphone and fed to an oscilloscope or a sound-level meter. The amplitude and frequency of the points of greatest response can be recorded manually. Better still, a "photostrip" can be made by pulling a film across the oscilloscope face at a speed synchronized with the sweep of the audio-frequency generator.

Once we had established the testing procedure we could address ourselves to

a question that had been worried for several hundred years and that had been answered in a number of different ways: What sounds should the top and back plates of an instrument produce before they are joined?

In 1840 Savart reported that "a top of spruce and a back of maple tuned alike produced an instrument with a bad, weak tone." He took the plates off a number of Stradivarius and Guarnerius violins (imagine!) and tested them, finding that the tap tones varied "between C sharp 3 and D3 (in the octave above middle C) for the top, and D3 and D sharp 3 for the back, always one tone or one semitone difference, the

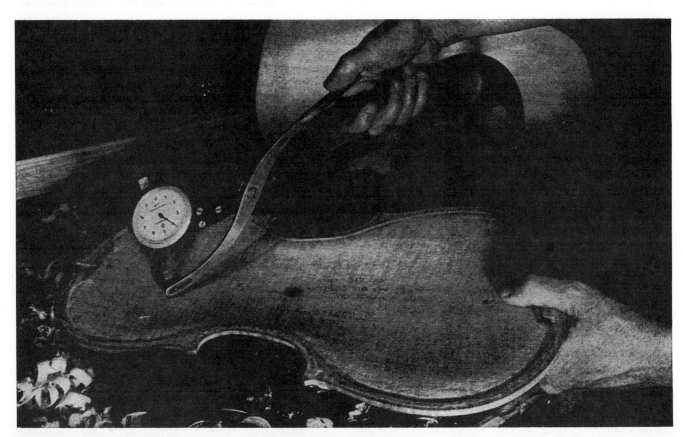

MEASURING PLATE THICKNESS makes it possible to determine where thinning can be done while maintaining a fairly uniform pattern of thickness. The measuring device, or caliper, consists of a dial gauge attached to one arm of an extended metal U.

back being higher than the top." Some violinmakers have held that the back should be a tone lower than the top; others, that the plates should be tuned to the same frequency.

My own findings are as follows: In the range of 120 to 600 cycles per second there may be one, two and possibly three peaks in the back plate and perhaps two or three more than that in the front. When the peaks of the front plate alternate with those of the back and the adjacent peaks are within about a semitone of one another, I get a good instrument. When the peaks coincide or are more than a tone apart, I get a bad one. Moreover, an average of the frequencies of the tap tones from front and back turns out to be just about seven semitones below the main wood frequency of the finished violin.

These conclusions are drawn from more than 400 photostrips of top and back plates of 35 instruments in the process of construction. After the plates were tested the instruments were assembled and then judged for tone quality by three criteria: (1) loudness test, (2) photostrips of the completed instrument and (3) actual playing by professionals. Then one plate, usually the top one, was removed and thinned, tested again and the instrument assembled for reappraisal. The back plate was thinned only when the top plate became so thin that it could no longer support string tension with safety. The entire thinning and testing process was sometimes repeated four or more times until each violin, viola or cello was judged to be good. So far I have spent six years on the program.

With our tap-tone test it is possible to follow the position of the main wood vibrations as they drift to lower frequencies when the wood is thinned and becomes more flexible. With a little practice one learns how to remove a few grams of wood from certain areas with a scraper or small plane and to estimate that the plate peaks (strong natural resonances) will move downscale, say 10 cycles per second. In some cases such a shift can make the difference between a good and a poor instrument.

As a kind of acid test of the theory I made a cello with the plate peaks matching; this is of course exactly wrong. During the next two years I gave the cello to several different cellists to play. All of them admired the workmanship and tried to be complimentary about the tone and playing qualities. The more forthright of them said that the tone was harsh and gritty in spots and weak in others and that the instru-

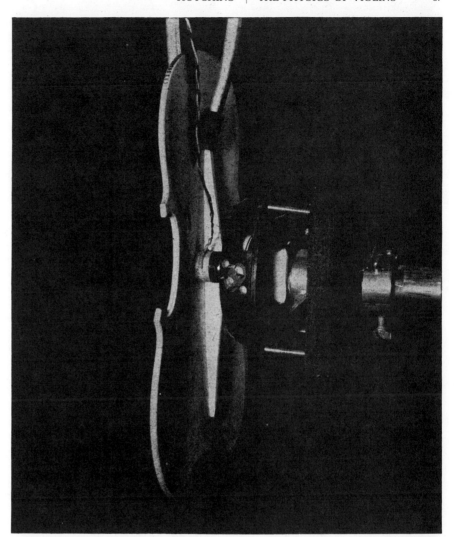

MAGNETIC DRIVER used in frequency-response tests is placed at the exact center of a top plate. The wires leading from the driver are connected to an audio-frequency generator that activates the driver over a range of frequencies from 20 to 20,000 cycles per second.

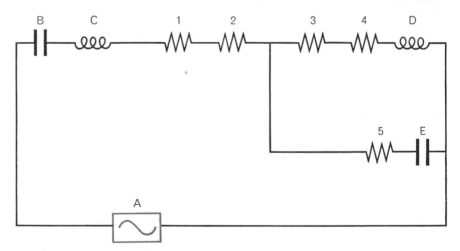

SIMPLIFIED ELECTRICAL CIRCUIT shows the nature of the two main resonances discussed in the text. The current from a constant-amplitude alternating-current generator (A) is analogous to the force applied by a given string to the bridge; this force is proportional to the string tension and amplitude of string vibration. The first capacitor (B) is analogous to a stiffness associated with the elasticity and dimensions of the wood; the first inductance coil (C) is analogous to a mass moving with the velocity of the bridge-string contact and having a kinetic energy equal to that in the wood. In instruments of the violin family the stiffness and mass of the wood largely determine its over-all response and the frequency of the main wood resonance. The main air resonance is determined largely by opposition of the air to compression when the f-holes are closed (E) and the mass of the air near the open f-holes (D). The five resistors in the circuit represent mechanical or acoustical resistances.

ment was particularly hard to play softly.

Finally I took the plates off, tested them again and removed about 10 grams of wood from the edges of the top plate so that the peaks of the top alternated with those of the back. In this condition Mischa Schneider played the cello in a concert by the Budapest String Quartet and pronounced it to be *magnifico*.

The greatest difficulty with the tap-tone test on a finished instrument is that both the top and back plates must be off at the same time so that they can be tested under the same conditions and without the complication of drift in the measuring equipment. The removal of both plates is a touchy operation even for an expert. With the help of several co-operative violinmakers, however, we have been able to test the plates of a few good old violins. More such tests are needed for definitive comparisons.

New Violins

It has been hundreds of years since the violin won its battle with the viols. The victory was not an unmitigated blessing. The variability of the shape of the viols, and particularly their flat back plate without a complicated set of resonances, meant that the instruments could be built in a variety of sizes that easily covered the entire range of pitch represented by the piano keyboard. On the other hand, the violin family leaves substantial gaps in coverage and, as has already been pointed out, its two deeper-voiced members do not have optimum musical characteristics.

Buried in private collections and museums there is a neglected but rich repertoire of polyphonic string music from the Renaissance period written for viols. Their characteristically thin and nasal, but uniform and distinctive, timbre blended well with the clavichord and cembalo, which were played at the courts of Renaissance nobles. Many of these gentlemen kept a chest of viols, usually consisting of six instruments, two each of the treble, alto and tenor sizes.

For contemporary performance of the viol repertoire, however, the old instruments are unsuitable. They do not have the variety of timbre that the violin has taught the modern ear to expect, and they do not have nearly the power to satisfy the requirements of a concert hall of even moderate size. On the other hand, the present family of violin, viola and cello have too much inequality in timbre and too great gaps in pitch to play the music as it was written.

The need for new instruments of the violin type has been considered by musicians and violinmakers for many years. The present work of developing the new instruments was initiated when Henry Brant, the composer-in-residence at Bennington College in Vermont, came to us with the problem. Brant felt that modern musicians, faced with the need to find an expressive language appropriate to the present day, are ever seeking to extend the powers of the bowed string instruments. The violin family remains the composer's most eloquent and expressive vehicle among all the instruments so far devised in Western music, but its members have been essentially unchanged for 200 years. More and more the need is being recognized for a gamut of graduated instruments of the violin type, with each member well enough developed to meet the test of solo as well as ensemble playing.

Changing the classical dimensions of the violin to create instruments of varied sizes and tunings has been tried many times without success. Now the necessary knowledge is at hand. I have indicated that the variables in the design of the violin are close to optimum. The object is to keep the two main resonances on the two open middle strings in spite of changes in size and tuning. It can readily be appreciated that it is no mean task to arrive at the correct proportions among physical variables—size, thickness and stiffness of wood, tightness of stringing and so on—that will produce the desired result in the resonance. For me it has meant years of literal cut and try, but with the help of scaling theory I am now close to having a set of empirical rules for making a genuinely complete family of instruments of the violin type. In doing this I have drawn heavily on the knowledge gained by other violinmakers who have tackled the same problem but without success because they did not have the benefit of modern acoustical physics. I have already built revised versions of the viola and cello, enlarging them somewhat to bring the resonances down to the frequencies of the open middle strings. As a result my viola has to have a peg at the bottom, like a cello, and is played between the knees. In addition I have added two new instruments to the family (one replaces the bass). This past January the six scaled members of the violin family were tried out at an informal concert, which a number of professional musicians found interesting and challenging as well as aesthetically pleasing. The smallest and the largest of the new instruments have not yet been finished and are giving the most trouble. Although scaling theory

tells us what to do, we are up against the limits of available materials and the human physique. For the smallest instrument, which is tuned an octave above the violin, material of sufficient tensile strength for strings is the major problem. Few materials have the strength to vibrate within the requisite range of frequencies and still provide strings long enough to allow the player to finger consecutive semitones. In the largest instrument the designer faces the mechanical problem of making it possible for the musician to bow and finger simultaneously.

Other violinmakers have experimented with instrument size. In the 19th century Jean Baptiste Vuillaume introduced a new model of the viola with an exceptionally large air volume, constructed on the scientific principles of Savart. He also developed a huge double bass, known as the *octobasse*, that was tuned by means of levers. Fred Dautrich of Torrington, Conn., spent much of his time during the 1920's and 1930's working on a graded series of instruments of the violin type that he called the vilonia, the vilon and the vilono. I have been fortunate enough to obtain a set of these. They are of such excellent workmanship and proportions that it has been possible to modify them slightly by applying scaling theory and adapt them to our present series of instruments.

In the past few years J. C. Schelleng, formerly of the Bell Telephone Laboratories, has been studying the violin as a circuit, one of the standard techniques of acoustics in which the various mechanically vibrating parts are treated in a manner analogous to the elements of an electrical circuit. Although the violin is exceedingly complicated, it possesses many simplicities not usually recognized. These, along with the fundamental physics of the instrument, permit the definition of "circuit elements" and lead to relations difficult to find empirically. This circuit concept is already being of great help in perfecting the new instruments, defining such problems as string tension, the mass of the box and the stiffness of the plates.

To sum up, I believe that, without ignoring the precious heritage of centuries, the violinmaker should become more conscious of the science of his instrument, and that the acoustical physicist should see that here is a real challenge to his discipline. We really ought to learn how to make consistently better instruments than the old masters did. If that challenge cannot be fulfilled, we should at the very least find out the reasons for our limitations.

7

The Physics of the Bowed String

by John C. Schelleng
January 1974

*What actually happens when a violin string is bowed?
Modern circuit concepts and an electromagnetic
method of observing string motion have stimulated
new interest in the question*

The heart of the violin or any of its relatives, the center from which flows the acoustic pulse that is the very life of the music, is the bowed string. The string—its action under the fingers and the bow, its gratifying responsiveness and even the problems it forces the player to solve—plays a major role in establishing the musical identity of this family of instruments. Conceptually the string is the simplest of components, although its manufacture calls for meticulous care: it must be flexible, uniform and strong. In spite of this

simplicity its action under the bow presents many unanswered questions. The elementary physics of its behavior can nonetheless be of considerable importance to the player.

Of the many papers published by Hermann von Helmholtz, ranging through physiology, anatomy, physics and the fine arts, there is one titled "On the Action of the Strings of a Violin" that appeared in the proceedings of the Glasgow Philosophical Society in 1860. Up to that time little was understood about what actually happens when a string is

bowed. Helmholtz' procedure is a good example of how a well-conceived experiment combined with simple mathematics can illuminate a problem that could not at the time be solved with either approach alone. Today we would call his apparatus an oscilloscope; to him it was a "vibration microscope," an invention he credited to the French physicist Jules Antoine Lissajous. Through this instrument he looked at a grain of starch fastened to a black string, which he set in vibration by bowing. The objective lens of the microscope was mounted on a

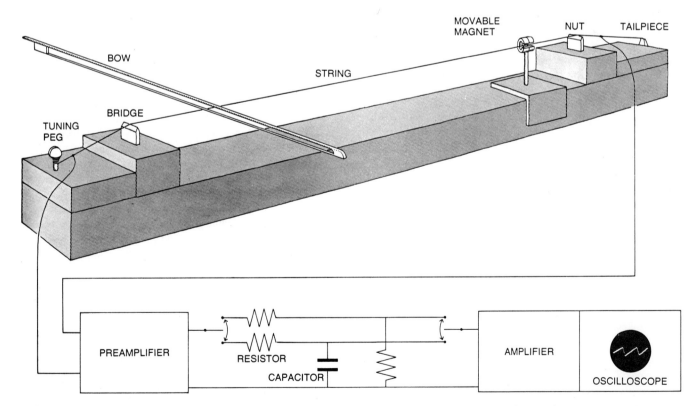

MONOCHORD, a simple experimental arrangement used by the author to study the motion of a bowed string, consists of a single electrically conducting string mounted between two massive bridges on a firm base. The movement of the string through the magnetic field set up by a small movable magnet generates an output signal that can be amplified and displayed on an oscilloscope screen (*see circuit diagram at bottom*). With the two switches in the up position the system displays string velocity; with the two switches down the system displays string displacement. The string can be bowed by hand, by a pendulum-driven bow or rotary bow.

large tuning fork so as to vibrate slowly parallel to the length of the string. When both string and fork were set in motion at suitable rates, Helmholtz saw a "Lissajous figure," a form of oscillogram that displayed the position of the starch particle as it varied within the period of vibration of the fork. By similarly examining the motion at other points he experimentally acquired the basis for a mathematical description of the motion of the string as a whole.

Helmholtz wrote that "during the greater part of each vibration the string is carried on by the bow. Then it suddenly detaches itself and rebounds, whereupon it is seized by other parts of the bow and again carried forward." Plotting the position of the bit of starch as a function of time, he found that every aspect of the picture as he found it except one could be represented by straight lines. During one period of vibration, almost regardless of where on the string he looked or where he bowed, the curve was a zigzag of two straight lines [see illustration below]. The two periods of time into which the vibration

was broken were always in the same ratio as the two lengths into which the point of observation divided the string.

Something can be learned merely by looking at the lowest string of an instrument while it is being vigorously bowed [see illustration on opposite page]. In appearance it widens into a ribbon bounded end to end by two smooth curves. (Actually the position around which the string vibrates is moved a little to one side by the average force exerted by the bow in its direction of motion.) Helmholtz found mathematically that the boundaries are parabolas; because of their shallowness they are indistinguishable from arcs of a circle. It would be a mistake, however, to suppose that the string itself has this shape at any time. The string, Helmholtz found, has at any instant the shape it would have if it were pulled aside by the finger to some point on the arc: it is a straight line sharply bent at one point. The bend races around this edge once in every vibration; for the open A string of a violin, for example, it goes around 440 times in one second. If Helmholtz had been able

to view the string with the aid of a stroboscopic lamp, the boundary would have disappeared and he would have seen the string as a sharply bent straight line. When the bow changes from "up bow" to "down bow," the motion of the bend around the edge changes from counterclockwise to clockwise.

The sideways velocity of the string at any point has two alternating values, unequal in magnitude and opposite in sign. As a result a typical zigzag displacement curve has a corresponding velocity curve that is rectangular in shape. The ratio of the two alternating velocities is the same as the ratio of the lengths into which the point of observation divides the string.

Two simple physical facts underlie the action of the bowed string. The first is that "sliding" friction is less than "static" friction and that change from one to the other is almost discontinuously abrupt. The second is that the flexible string in tension has a succession of natural modes of vibration whose frequencies are almost exact whole-number multiples of the lowest frequency; as a result the duration of a single vibration in the first, or lowest, mode precisely equals the duration of two vibrations in the second mode, three in the third and so on. Without outside compulsion the string is therefore by its very nature given to supporting a "periodic" wave, that is, a repetitive series of similar vibrations with a wave form dictated by the "stick-slip" process. The string allows the coexistence of a multitude of harmonics; the peculiarities in friction require it.

Helmholtz' shuttling discontinuity is the timekeeper that precisely triggers the capture and release of the string at the bow. There is a perennial explanation that views the string as a spring periodically pulled sideways to the breaking point of static friction. The spring recoils and is again captured. This view cannot explain the constancy in repetition rate over the wide range of bow forces, or "pressures," applied by the hand. The correct explanation must be given in dynamic terms; such an explanation is suggested by the constancy in time needed in the flexible string for the bend to travel twice its length. The timing is vividly illustrated by striking a long, taut clothesline near one end with a stick. A discontinuity is clearly seen hurtling to the far end, where it is reflected. On its return one feels through the stick (still resting on the line) an impulse much like the momentary frictional force on the string, which fails to hold at escape but succeeds at recapture.

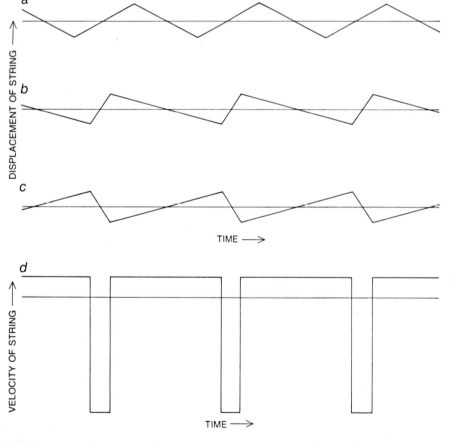

DISPLACEMENT OF BOWED STRING from its average position is plotted as a function of time in the first three curves in this illustration. The characteristically zigzag curves were obtained by bowing near one end of the string and observing at the middle (a), near the bridge end (b) and near the nut end (c). In each case the two periods of time into which the vibration is broken are in the same ratio as the two lengths into which the point of observation divided the string. Rectangular string-velocity curve (d) corresponds to curve c.

A simple experiment partially confirms this picture of the action of the bowed string [*see illustration on next page*]. An instrument is mounted with the strings horizontal. A light bow, suspended at its heavy end by a long thread, rests on one of the strings at a point near the bridge. A second bow sets the string in vigorous vibration. Before the hanging bow can begin to move slipping occurs at all times except at moments when the string reverses. Since friction in slipping is nearly independent of speed of slipping, the forces in the two directions of vibration are the same but the impulses imparted to the bow are proportional to the duration. The direction of the acceleration of the hanging bow will therefore indicate the direction of string motion during the longer duration.

The experiment shows, however, that the direction in which the hanging bow moves is the same as the direction in which the driving bow is moving. Therefore it is in the longer interval that the string moves with the driving bow; relative motion between driver and string is accordingly less than in the shorter interval and sticking is presumably occurring. If the hanging bow is now placed near the opposite end of the string, it will be found to move in a direction opposite to that of the driver.

Helmholtz believed that the velocity of the string while it is snapping back is constant. A half-century later C. V. Raman found that in most cases this is only approximately true. Raman's discovery came in the course of an ingenious study of the mechanical action of the violin involving both experiment and theory. With respect to the bowed string his point of departure was to describe the motion in terms of the progressive waves of transverse velocity that make up the standing waves of the Helmholtz system. The same wave can be described in terms of its lateral displacement or its lateral velocity. One advantage in emphasizing velocity is that these waves can be represented as straight lines.

The shape of the Raman wave in those cases that are of interest in music (Raman dealt with many that are not) again is a zigzag, but it differs from the displacement curves introduced above in that although the "zigs" are slow, the "zags" are instantaneous [*see top illustration on page 73*]. When such a wave is reflected from the immovable end of the string, it looks exactly as it did before except that its direction of propaga-

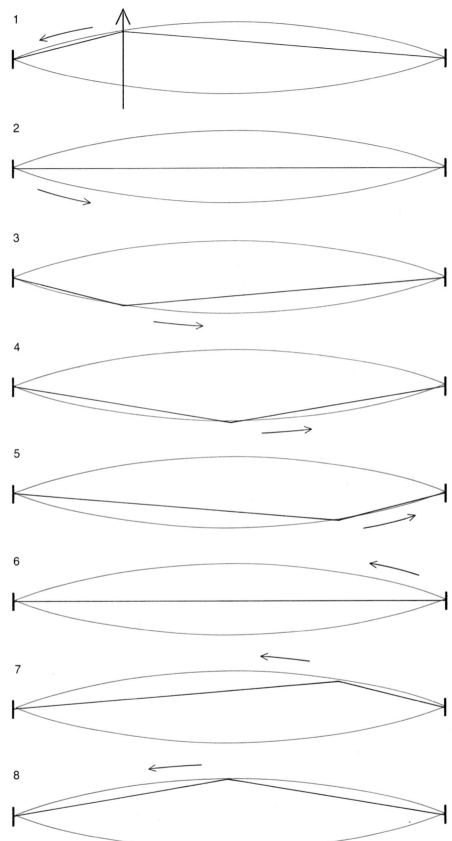

SHAPE OF BOWED STRING appears to widen into a ribbon bounded end to end by two smooth parabolic curves (*colored outlines*). As Hermann von Helmholtz found more than a century ago, however, the actual shape of the string at any instant is a straight line sharply bent at one point (*black line*). The bend races around the boundary once in every vibration. The direction of the bend's circulation in this particular series of diagrams corresponds to an upward motion of the bow; reverse the direction of the bow's motion and the bend reverses its direction of circulation. This peculiar motion is a form of standing wave.

tion is reversed. When vibration is in the fundamental mode, the length of the string is half the distance between zags.

This progressive velocity wave is of interest because, being incident on the bridge of the violin, it exerts a vibrational force whose shape is identical with its own. Insofar as the Helmholtz approximation holds, its harmonic structure therefore describes the tone quality of the string itself at the point in the instrument where the sound spectrum has not yet been influenced by resonances in or radiation from the body. The spectrum is remarkably simple: the amplitude of the nth harmonic is $1/n$ times the amplitude of its lowest, or fundamental, frequency. This simple relation is of considerable importance in investigating the sound spectrum of the violin as a whole.

Within the past few years, stimulated by circuit concepts and an electromagnetic method of observing string motion, there has been a renewal of interest in the physics of the bowed string both in this country and in Europe. More than half of the strings currently in use in stringed instruments are electrically conductive. If a small magnet is placed close to the string, the combination of a conductor moving in a magnetic field constitutes a magneto whose output can be displayed merely by inserting the string in the input circuit of a suitable amplifier connected to an oscilloscope. The electromotive force is proportional to the velocity of the string. The string can be mounted on the instrument proper or on a monochord: an experimental arrangement consisting of two massive bridges on a firm base with some means for providing tension in the

string and a mount for the magnet (*see illustration on page 69*). In my experiments two methods of bowing were used besides bowing by hand: a rotary bow developed by F. A. Saunders in his researches on violins and string action, and an ordinary bow driven by a 50-pound pendulum.

An electrical circuit connected to the monochord (or to the instrument) makes it possible to display the velocity or displacement of the string in the form of oscillograms. The first oscillogram in the bottom illustration on the opposite page, for example, displays velocity at the bow in a very flexible string. In the long period velocity lies above the zero line; that is also the velocity of the bow (if one ignores the slight ripples). In the short period of slipping there is a high negative velocity as the string whips

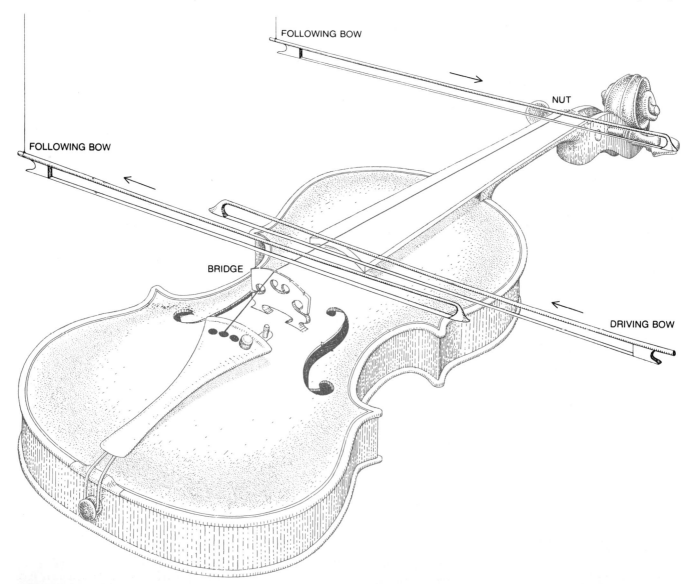

"FOLLOWING BOW" EXPERIMENT, performed in the course of the author's investigation, partially confirms Helmholtz' dynamic picture of the action of a bowed string. With the instrument mounted horizontally, a light bow, suspended at its heavy end by a long thread, rests on one of the strings at a point near the bridge. A second bow sets the string in vigorous vibration. In this situation the hanging bow is found (after a short period of slipping) to move in the same direction as the driving bow. The direction in which the hanging bow moves indicates the direction of string motion during the longer interval of each vibration. When the hanging bow is placed near the opposite end of the string, one finds that the "follower" moves in a direction opposite to that of the "driver."

backward to take a new hold on the bow. The bow was near the bridge in this case and the shape of the curve is close to what the Helmholtz construction predicts. The zigzag in the second oscillogram shows the same vibration in terms of displacement instead of velocity.

Simplicity in instrumentation is not the only advantage in displaying velocity rather than displacement. In this way high-frequency detail that would not be suspected is brought out clearly.

When the circuit is arranged to indicate displacement close to the bridge, it also indicates the transverse vibrational force applied by the string. The "sound of strings" alone, divorced from the modifying effect of the body of the instrument, can be produced by placing a magnet for each of the strings close to the bridge. The output from the four magnets in series is then passed through the integrator and is amplified and recorded. This arrangement sums up the forces applied by all strings. Playback therefore gives the effect of strings alone. (Recording is unnecessary if a dummy fiddle radiating no sound is used.)

The result definitely resembles a bowed-string instrument, but an inferior one. If the system faithfully translates the force on the bridge into radiated sound pressure, the resulting sound spectrum with a given speed of bow will vary with frequency in an essentially inverse manner, the strongest effect being in the lowest tone. This "sound of strings" is not unlike the sound of the violin family in its upper octaves, but it differs drastically in the lower ones, where instead of the fiddle's achieving its strongest fundamental on the lowest tone the effect falls practically to zero because of the small size of the instrument in relation to the wavelength of such a tone in air. The harmonics then are to be credited with the fundamental tone heard subjectively. It is hardly necessary to emphasize the role that this characteristic has played in the evolution of the instrument.

In playing a bowed-string instrument there are certain limits to the speed of the bow, its distance from the bridge and the normal force applied; these limits must not be overstepped in a specific musical situation. For the experienced player this is usually a matter of choosing almost subconsciously from familiar patterns of action, but in extreme cases he is probably always aware of the limitations imposed on him. The ranges of these mechanical parameters are fortunately wide: the bow-to-bridge distance, for example, may vary from a

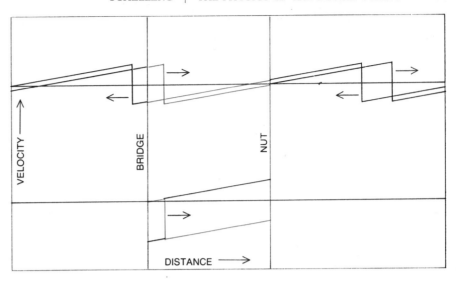

RAMAN WAVES were introduced by C. V. Raman to describe the motion of the bowed string. The shape of such a progressive transverse velocity wave differs from the corresponding Helmholtz standing wave of string displacement in that whereas the "zigs" are slow, the "zags" are instantaneous. When the oppositely moving Raman waves (*top*) are summed, the resulting wave (*bottom*) shows how the two velocities that exist alternately at any point on the string depend on the position of the discontinuity between "slipping" and "sticking."

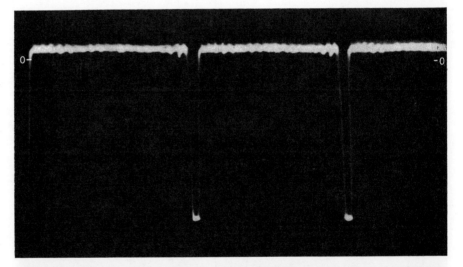

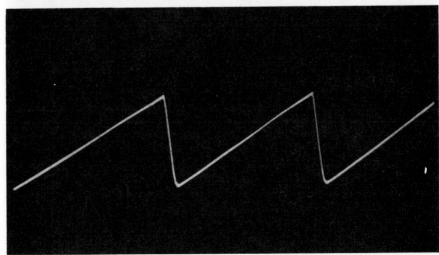

MOTION OF VERY FLEXIBLE STRING at the bow is represented by these two oscillograms, which show string velocity (*top*) and string displacement (*bottom*) for the same vibration. The bow in this case was located about a twentieth of the length of the string away from the bridge. The shape of the curves is close to what Helmholtz picture predicts.

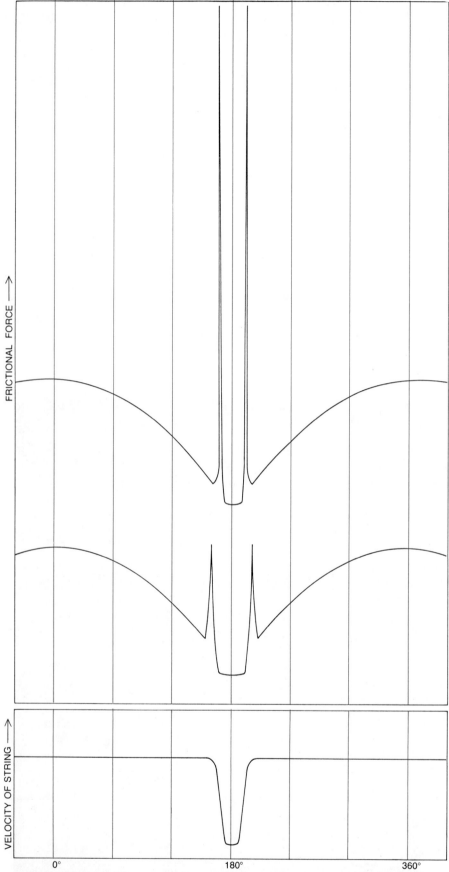

FRICTIONAL FORCE ⟶

VELOCITY OF STRING ⟶

0°   180°   360°

"RABBIT EARS" characterize the curves that represent the frictional force that is exerted
by a bow in order to vibrate a string. One ear forms on the curve for frictional force when
the discontinuity arrives from the nut, overcomes static friction and initiates slipping; the
other ear forms when the discontinuity returns from the bridge and initiates sticking. The
top curve shows the frictional force when the bow force has a typical value midway between
the upper and lower allowable limits; the middle curve applies to the lower limit. For clar-
ity ripples have been omitted from the wings of both of these curves (left and right). The
curve at bottom is the string-velocity curve corresponding to the frictional-force curve at top.

minimum value to five times that minimum; speed and force may range up to
100 times their minimum value. Given
any two of these parameters, in order to
assume an acceptable tone the third
must fall within a range that depends
on the physical constants of the string
and the body of the instrument. For sustained tones these ranges are generous,
although clearly all portions of a given
range are not equally desirable. For example, position and speed being given,
the highest permissible force may typically be 10 times the minimum. The first
question is: What are the processes that
determine the existence of these limits?

In order to explain the limitations on
bow force it is helpful to consider how
the frictional force at the point of contact of the bow and the string varies in
time. Although one cannot at present display this force on an oscilloscope, it is
possible to form a simplified qualitative
picture on the basis of the physics involved. To do this one assumes (1) that
the elementary laws of static and dynamic friction can be applied, (2) that the
bridge acts as a high "mechanical" resistance (analogous to resistance in an
electrical circuit) and (3) that the mass
and tension of the string and its motion
at the bow are known.

Assuming that the force on the string
is always in the direction of the bow's
motion, the points of maximum bow
force occur at intervals of zero degrees,
360 degrees, 720 degrees and so on;
the minimum points will then fall at 180
degrees, 540 degrees, 900 degrees and so
on [see illustration at left]. The cyclic
swinging of force between these two
levels is what is needed in order to vibrate the instrument. In the small interval around 180 degrees slipping occurs.
In transition from sticking to slipping
there is a brief moment when the maximum of static frictional force required
by bow force is exerted. Research has
shown that "static friction" is really not
quite static: velocity, although minus-
cule, is finite, and friction in reality
changes continuously with velocity near
zero velocity with a narrow maximum
corresponding to static friction. The
same curve is presumably followed in reverse in going from slipping to sticking.
The "rabbit ears" evident in such frictional-force curves are the result. Such
curves differ in detail in going from note
to note because of complexity in the action of the body.

Consider the situation where the bow
force has a typically intermediate value.
From the zero-degree mark onward, with
the string clinging to the bow, the force
falls toward the minimum, which is dic-

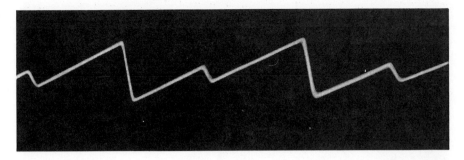

NEW ZIGZAG begins to form in the oscillographic displacement curve of a bowed string when the bow force is allowed to fall below the minimum bow force. If this unstable condition is allowed to develop, the fundamental tone will soon give way to the octave tone.

tated by dynamic friction. Then, at the moment when sticking seems most secure, the discontinuity arrives and overcomes static friction. The discontinuity needs to provide only the amount by which static friction exceeds dynamic friction. The discontinuity is capable of contributing more than is required here, perhaps much more. As the bow force is increased, however, the time will come when the discontinuity will lose out in this test of strength and the vibration will become an erratic squawk. Maximum bow force will have been exceeded.

A different kind of failure occurs when the bow force is decreased to a minimum. Here the "ears" of the frictional-force curve fall to the same level as the maximum force at the zero-degree mark, and with the lightest additional decrease static friction (as indicated by the "ears") becomes insufficient to hold the string near the zero-degree level. The result is an unstable string-displacement curve in which a new zigzag begins to form [see top illustration on this page]. If this new zigzag is allowed to develop, the fundamental tone will be replaced by the octave tone; in short, one will have failed to provide minimum bow force.

It is an important fact in the mechanics of playing that maximum bow force, which depends primarily on the string and on frictional coefficients, is inversely proportional to the first power of the distance of the bow from the bridge, whereas minimum bow force, which in addition depends on the body of the instrument, is (at least approximately) inversely proportional to the second power of the same distance. The quantities necessary for calculating these limits are known well enough to explain how the string reacts to bow forces. For sustained tones with a given bow velocity one can display the logarithmically linear trends of maximum and minimum bow force in terms of the relative distance of the bow from the bridge expressed as a fraction of the total length of string [see bottom illustration on this page]. The most important result is that the maximum bow force and the minimum bow force are equal when the bow is placed at a certain point very close to the bridge and that they diverge as the bow moves away from it. (Actually the curves near the intersection are to be regarded as extrapolations from the right, where normal conditions obtain.) It is this open space between the limits that gives to bow force the wide tolerance that makes fiddle-playing possible.

The forces toward the left between these lines are impractically high; nor-

mal playing is confined to the area toward the right. Farthest from the bridge the volume of the sound is least, the content of upper harmonics is minimal and the timbre has the gentle character that composers seek by designating *sul tasto:* "bow over the fingerboard." Exceed maximum bow force and the result is unmusical; fall short of the minimum and the solid fundamental tone is lost, leaving what is sometimes called a surface tone. The closer the bow is to the bridge, the less generous is the ratio between maximum and minimum bow force and the steadier is the hand that is needed. The experienced player prizes

this domain for its nobility of tone; the beginner finds it prudent to play closer to the fingerboard. Closer still to the bridge bow force mounts prohibitively and the solidity of the fundamental tone disappears until little more than a swarm of high harmonics remains to suggest the fundamental tone; this is the eerie *sul ponticello* ("bow over the little bridge") of the composer. Within the normal-playing area the relative harmonic content increases—the tone becomes more brilliant—either as the bow moves toward the bridge or as the bow force increases toward its maximum.

Such a diagram is to be taken as quali-

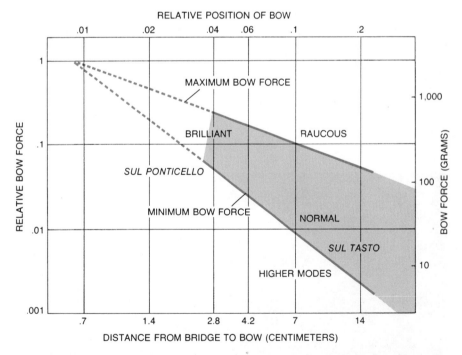

NORMAL-PLAYING RANGE for a bowed-string musical instrument is depicted for sustained tones at a constant bow velocity in this graph, which indicates the logarithmically linear trends of maximum and minimum bow force in terms of the relative distance of the bow from the bridge expressed as a fraction of the total length of the string. As the graph indicates, the maximum bow force and the minimum bow force tend toward equality (*upper left*) when the bow is placed at a certain point very close to the bridge and they diverge as the bow moves away from the bridge (*lower right*). The open space between these two limits, shown in color, accounts for the wide tolerance in permissible bow force. *Sul tasto* means "bow over the fingerboard"; *sul ponticello* means "bow over the little bridge." The second set of coordinates (*bottom scale; right scale*) suggests the normal-playing conditions for a typical A string of a cello bowed at a sustained velocity of 20 centimeters per second.

tative, in particular the curve for minimum force, which varies greatly from note to note because of the complexity in the response of the body of the instrument. Although the Helmholtz idealization is close enough to fact to provide a useful basis for many first-order calculations such as those described above, it is not completely trustworthy in other respects. In contradiction to what it implies, harmonic content increases with bow force, changing timbre and loudness. If loudness depended only on the "root mean square" vibrational force on the bridge, the effect would not be of much consequence, but when harmonics are radiated more efficiently than the fundamental tone or perceived more sensitively by the ear, the effect is of some importance. The fact remains that the player's major resource in controlling volume lies in the speed and placement of the bow. The implication that sound pressure is directly proportional to bow speed and inversely proportional to bow-to-bridge distance is not far wrong.

Bowed tones start in different ways, but perhaps most frequently the bow pulls the string sideways until its displacement can no longer be supported by static friction. Failure in friction, like release in plucking, sets up two oppositely traveling Helmholtz discontinuities, only one of which, the one toward the bridge, can become sustained. Until bow speed matches bow force, however, the condition is described as "raucous," and there may be many false starts before balance is realized. The art in such beginnings is to achieve the match in such a short time or at such a low sound level that an unpleasant effect is avoided.

A noiseless beginning can be made by allowing the bow already in motion to make a "soft landing" on the string, thus entering the normal-playing zone by way of the zone labeled "Higher modes." Theoretically at least, bow forces and velocity can be balanced from the beginning.

In the foregoing discussion of the frictional force between the bow and the string a phenomenon of some interest was left unmentioned in the interest of simplicity, namely the role of the "ears" of the frictional-force curve in setting up reverberations between the bow and the ends of the string, some of which may persist for many periods of vibration. These effects are ignored in the classical discussion of the action of the bow but are prominent in oscillograms of string velocity. Consider a curve showing the motion of the string under the bow, where during the long interval of sticking the string might be expected to follow the unchanging speed of the bow [see illustration below]. It is in fact true that there is no slipping, but the string can nevertheless move by rolling on the bow except as prevented by the string's resistance to twist. The ripple in rolling implies a corresponding ripple in force exerted on the string.

A more comprehensive frictional-force curve would therefore show sharp fluctuations in force superposed on the smooth sections. One effect is to raise the minimum bow force somewhat. Away from the bow the ripples in velocity are much more pronounced. The motion would be completely suppressed at the bow if the string allowed no twisting,

but it would still exist away from the bow.

The term wolf note is commonly used to describe an unpleasant sound that appears consistently at an isolated frequency in a bowed-string instrument. Often its origin is obscure. There are many varieties of wolves but the most vicious of the species has its habitat in the cello (and sometimes in the violin or the viola) one octave and a few semitones above the lowest note. There is no mystery about its cause. The body of each instrument has a multitude of resonances, and the wolf tone (if one exists) arises at the most prominent of them. For the bowed string to behave properly its ends must be given a support whose rigidity is in keeping with the mass of the string. Fingering one of the heavier strings to the same frequency as the resonant frequency of the body therefore invites trouble. This nuisance manifests itself in different ways but the most characteristic way is the generation of two tones, both forced vibrations, close enough together to produce a harsh beating. Since the two tones straddle the resonance peak, they require less bow force than one tone would alone at the resonant frequency [see top illustration on opposite page].

In the stiff strings of the piano, consisting as they do of thick steel wire, the frequencies of higher modes of vibration are not whole-number multiples of the frequency of the lowest mode but are somewhat sharp. That is not a defect: the "tang" it gives to the sound of the struck string is highly valued. What effect does stiffness produce in a bowed string? Clearly it is different from that in the piano. The mechanism of bowing produces a succession of almost identical vibrations. From a mathematical point of view this is another way of saying that the vibration is made up of harmonic components whose frequencies are exact whole-number multiples of the lowest frequency.

There can indeed be an effect in the bowed string. Although inharmonicity is perforce held at bay, freedom in vibration is restricted. One expects deterioration in tone quality through reduction in higher harmonics, difficulties in intonation and the need for abnormal bow forces. Before 1700, when wound strings became available, all violin strings consisted of gut, but the gut G string (the lowest) was unsatisfactory. The reason is not hard to find. With bowing the fundamental frequency is close to that of the lowest natural mode of vibration; in the gut G, however, seven times that frequency falls midway between the

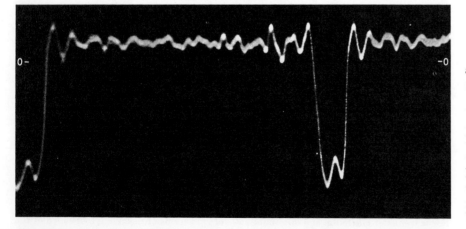

ROLLING OF STRING on the bow during a long interval of sticking produces characteristic "ripples" in the string's velocity curve. To obtain this oscillogram an A string of a cello was bowed with 4.5 times the minimum bow force. The period immediately following the capture of the string by the bow (that is, the section of curve just to the right of each main pulse) shows mainly the decay of the pulse formed at that capture as it reverberates in the short section of string. The period before release shows mainly long-delayed reverberations in the long section, not only from the most recent release but also from earlier ones.

sixth and seventh vibrational modes and so is completely without the support of resonance. Regardless of the bow force, the seventh harmonic must be negligible. This difficulty in producing harmonics can be illustrated by a curve that shows velocity at the bow in a stiff string [*see bottom illustration at right*]. When such a string is bowed with minimum bow force, its behavior bears almost no resemblance to that of a flexible string.

The sharpening of the *n*th mode caused by stiffness is directly proportional to the square of *n* and to a quantity called the coefficient of inharmonicity. If one changes a string mounted at a given position on a given instrument, keeping length and frequency unchanged, interchangeability requires that the same tension be maintained as well. Considering now a series of homogeneous strings of diverse materials, the coefficient of inharmonicity turns out to be proportional to the modulus of elasticity divided by the square of the density. For steel this ratio is about 50 percent greater than it is for gut; for aluminum the ratio is nearly five times greater than it is for gut. For silver, on the other hand, the ratio is about a third of the value for gut. A steel piano string that has the same pitch as the steel *E* string of the violin has an inharmonicity coefficient 20 times greater. If one ignores differences in tension in the four strings of a violin (actually it is considerably more in the highest string than in the others), inharmonicity for homogeneous strings of the same material, being inversely proportional to the fourth power of frequency, increases by a factor greater than 100 in going from the highest string to the lowest.

From calculations and measurements for several strings on the market, including some that are wound, it appears that when the coefficient of inharmonicity is equal to or less than .1, stiffness offers no disadvantage in bowing. In the steel *E* string of the violin the value is about .04. For one very good cello *C* string with metal winding on steel cable it is about the same.

As with many seemingly simple things, there is much about the bowed string that remains open to speculation. Does twisting lead to any important acoustic effect? How important is "negative resistance" throughout slipping and exactly how does rosin behave? How much do successive periods of vibration differ and is this slight wandering of any musical significance? The answers to such questions may be of little interest to the player, but the student of the bowed string would still like to know.

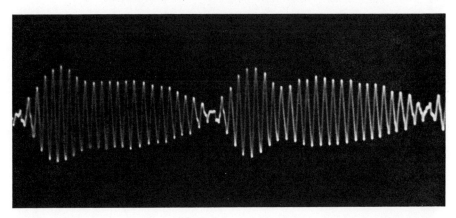

"WOLF NOTE," an unpleasant sound that may appear consistently at an isolated frequency in a bowed-string instrument (particularly in the cello), is produced by the "beating" of two or more tones generated by "forced" vibrations of a string clustering around the natural resonant frequencies of the body of the instrument. This oscillogram, which shows the string motion for a complicated wolf tone obtained from the *C* string of a cello, was supplied by Ian M. Firth and J. Michael Buchanan of the University of St. Andrews in Scotland.

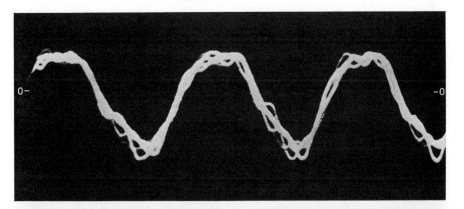

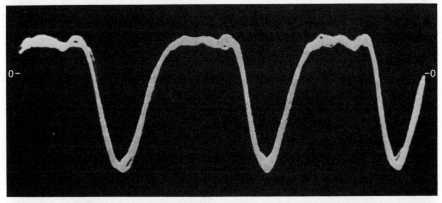

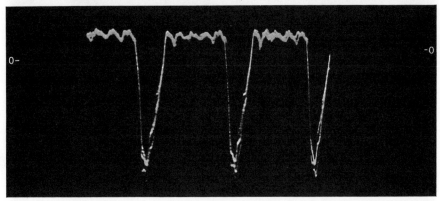

EFFECT OF STIFFNESS in a bowed string is a deterioration in tone quality attributable in part to an increased difficulty in producing higher harmonics. This problem is demonstrated by these three oscillograms, which represent the velocity of a stiff string at the bow for three levels of bow force: minimum (*top*), intermediate (*middle*) and high (*bottom*).

8

Architectural Acoustics

by Vern O. Knudsen
November 1963

*Sound is as much a part of man's man-made
environment as heat or light. It can now be effectively
managed, notably in rooms where music is heard, by
applying the principles of acoustical physics*

The opening of a large concert hall these days is almost inevitably followed by a spate of reports, reviews, criticisms and opinions about its acoustical qualities. Amateurs and competent critics alike try to compare the music heard in the new hall with their recollection of the same or similar music heard in concert halls of acknowledged acoustical excellence. This exercise in auditory memory is not easy, and it gives rise to many pretentious statements. Yet the fact remains that, as in winetasting, the subjective evaluation of experts is the court of last appeal. For this reason architectural acoustics is an art as well as a science. If a new concert hall shows palpable deficiencies, as sometimes happens even today, the impression is strengthened that the science of acoustics has failed, or at least has been found wanting. Such a judgment is much too harsh. What usually happens in such cases is that the available knowledge, for a whole complex of reasons, has not been adequately applied. For example, critics reported serious deficiencies in the acoustics of Philharmonic Hall in New York, which opened a year ago. The original acoustical engineers and an independent team of consultants have now established these deficiencies by objective methods, and a number of changes have been made in the auditorium's design. The results of these changes, however, are still subject to critical evaluation, and it would be premature to discuss them here.

The purpose of this article is to describe the objective acoustical elements that have led to the design of many fine music halls and auditoriums. The application of acoustical knowledge to architecture dates back barely 60 years. Until about 1900 the design of a successful music room was almost entirely a matter of luck. Today the design can

be based on well-established principles of physics and engineering.

Acoustics is one of the oldest branches of physics. It originated in the study of music, which probably began with Pythagoras more than 2,500 years ago. By means of a single stretched string he showed that consonant intervals in music can be expressed by ratios of simple whole numbers. Acoustics has come a long way since then, both as an independent branch of physics (physical acoustics) and in association with other sciences and arts. In the second category are psychoacoustics and physiological acoustics, which deal broadly with the nature of speech and hearing; communication acoustics, which deals with the auditory aspects of telephony, radio and sound reproduction; musical acoustics, which deals with the acoustics of the human voice and musical instruments, and architectural acoustics.

Acoustics first became associated with

REFLECTION AND DIFFRACTION of sound waves can be studied by photographing the wave patterns created by an electric spark. This sequence shows the waves generated by a spark in a model of Royce Auditorium at the University of California at Los Angeles. The spark wave originates from a point on the stage and travels to the rear of the auditorium,

architecture when men began to assemble in groups to hear speeches, listen to music and see and hear plays. To create a favorable setting for such activities the Greek and Roman open-air theaters and forums evolved, and many of them have survived to this day. The typical open-air amphitheater consists of steeply banked benches arranged in a semicircle in front of a platform. With the passage of time the platform evolved into a stage with massive rear and side walls of masonry (and sometimes a ceiling) that served the acoustical purpose of reflecting, directing and thereby reinforcing the sound intended for the audience. Vitruvius, the first-century Roman architect and engineer, wrote that large vases tuned as resonators were often located in the seating area to reinforce certain sounds. Whether or not such vases were actually used is uncertain, but in any case they could only have absorbed sound, not reinforced it.

The Greeks did, however, develop one acoustical device of considerable value: the masks worn by actors. In addition to providing exaggerated facial expressions appropriate to the various roles, the masks served as megaphones that improved the mechanical coupling between the voice-generating mechanism and the surrounding air. A megaphone does not amplify the voice, but it does enable more of the available vocal energy to emerge in the form of sound waves than would emerge without the aid of the megaphone.

The principal defect of the Greek and Roman theaters is that the semicircular tiers of seats act as reflectors that tend to focus sounds from the stage back to a point on or near the stage. Moreover, the echoes from concentric tiers are reinforced at certain frequencies and diminished at others. The reason is that the vertical risers, which form the backs of the benches, create an echelon of uniformly spaced reflecting surfaces. The reflected waves are in phase and reinforce each other when the distance between risers is equal to one, two, three or any other whole number of half-wavelengths. When the distance between risers is one, three, five or any other odd number of *quarter*-wavelengths, the reflected waves meet in contrary phase and thus tend to cancel each other [*see illustration on page 82*]. For example, risers that have a spacing of 2.5 feet will constructively reinforce a series of sounds that have wavelengths in feet of 5, 2.5, 1.67, 1.25, 1 and so on, corresponding to tones that have frequencies in cycles per second of 225, 450, 675, 900, 1,125 and so forth. These frequencies constitute a harmonic series. The same riser spacing of 2.5 feet leads to wave cancellation in a series of odd-numbered harmonics with frequencies of 112.5, 337.5, 562.5, 787.5 and so on.

The effect of such wave reinforcement and cancellation can readily be demonstrated by speaking, singing or clapping hands on the stage of a typical Greek or Roman open-air theater. The sound reflected from the tiers of benches produces a sustained echo whose characteristic pitch is determined by the distance separating adjacent risers. As a result, when speech or music is heard in an open-air theater—or in a room or auditorium in which there are parallel and uniformly placed reflecting surfaces—the reflected sound may suffer a serious distortion in frequency. Fortunately in an open-air theater these frequency-dependent reflections generally pass over the heads of the audience, but since the reflections come to a focus on the stage they can be extremely disturbing to performers rehearsing in an empty theater. The problem largely disappears, how-

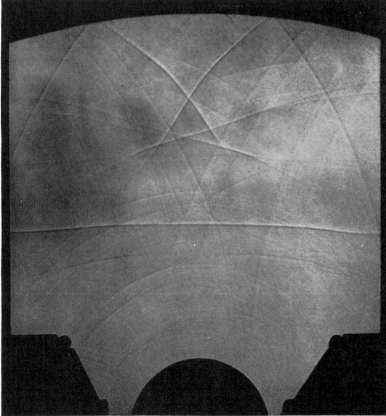

shown in plan view. (The black semicircle is not part of the plan but a mask to keep the bright spark from fogging the film.) The wave front begins as a simple arc (*left*) and becomes almost straight as it is reflected from the concave rear wall (*middle and right*). This produces an echo on the stage and in the front rows of seats. The echo was reduced by treating the rear wall with absorptive material. Note the complex patterns created by the proscenium. The photographs are by L. P. Delsasso and the author.

SIMPLE PLOTTING OF WAVE REFLECTIONS, called ray acoustics, has only limited value for predicting the acoustics of an auditorium. Several rays are shown superimposed on a spark photograph of Royce Auditorium. Note that rays *A* and *B* on reflection (*A′*, *B′*) fail to predict the complex diffraction patterns from the proscenium at left of stage. Rays *C* and *D*, however, represent reasonably well the reflections from a straight wall.

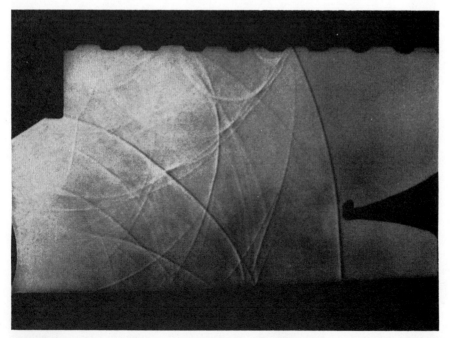

MODEL OF LONGITUDINAL SECTION of Royce Auditorium produced this complex wave pattern in a spark photograph. Diffusive reflections from the coffered ceiling cannot be predicted by ray acoustics. Analysis of such patterns is the objective of wave acoustics.

ever, when a capacity audience provides a sound-absorbing covering on the tiers of benches.

New acoustical problems arose when civilization and culture spread northward and it became necessary to provide enclosed buildings for theaters, churches and other auditoriums. In these buildings sound echoed from walls and ceilings, and when the enclosed space was finished with hard and sound-reflecting materials such as marble, stone and concrete, the architect encountered a vexing problem: excessive reverberation. This phenomenon is merely sustained echoing, and for the most part it was accepted as an inevitable result of building large enclosures of durable materials. Indeed, much of the majesty of a cathedral derives from the sonorous reverberations it imparts to voices and musical sounds. Although such reverberations may be acceptable, and even desirable, for church music and services, they must be held to strict limits in designing a lecture room, an auditorium, a concert hall or an opera house.

Acoustics of Closed Spaces

In order to handle the acoustical problems of enclosed spaces architects, and more recently acoustical engineers, have developed two basic procedures. The earliest and simplest utilizes ray, or geometrical, acoustics; the more recent and comprehensive procedure requires a detailed analysis of how waves of different frequencies actually interact with reflecting and absorbing surfaces of various shapes and dimensions.

Ray acoustics assumes that sound waves travel in straight lines and that when they encounter a new medium, such as the wall of a room or any substance whose density or elasticity differs from that in which the sound originated, the waves are reflected, refracted and transmitted in a fashion that is uniform for all wavelengths. It is assumed, for example, that sound waves are reflected from surfaces in the same way a billiard ball without spin rebounds from a cushion; in other words, that the angle of reflection equals the angle of incidence. In employing ray acoustics architects superimpose on their two-dimensional plans and sections families of straight lines that represent incident and reflected sound waves. This simple technique is useful for uncovering gross acoustical faults such as focusing effects from concave surfaces and for determining the shape of enclosures that will give optimum distribution of sound.

Ray acoustics is valid, however, only

for wavelengths that are small compared with the dimensions of the reflecting surfaces. The wavelength of a sound wave that has a frequency of 1,000 cycles per second is about 1.1 feet. For such a sound wave to be reflected in the simple manner predicted by ray acoustics, the dimensions of the reflecting surface must be at least two or three times the wavelength, or two or three feet. For wavelengths that are not short compared with the dimensions of the reflecting surfaces, ray acoustics fails. Thus sound that has a frequency of 100 cycles per second and a wavelength of 11.25 feet will not be reflected in a simple manner from a surface that measures two or three feet across. The limitations of ray acoustics become apparent when one considers that the wavelength of audible sound varies from about 56 feet at the low-pitch end of the range to less

than an inch at the high-pitch end. Virtually every architectural detail in an auditorium or music room will be large compared with the shortest wavelengths and small compared with the longest.

In order to overcome the limitation of ray acoustics one must employ wave, or physical, acoustics, which is based on the physical theory of waves. Only wave theory can cope with the real behavior of sound in rooms. The preceding discussion of reinforcement and cancellation in sound waves reflected from the tiers of benches in an open-air theater gives an example of wave acoustics. An analysis of an open-air theater by means of ray acoustics would show only that sound reflected from the curved tiers of benches would come to a focus at the stage. Such acoustical phenomena as room resonance, reverberation at low frequencies, interference, diffraction and

the reflection and transmission characteristics of openings in a room or of systems of suspended panels can be understood and subjected to control only by the rigorous application of wave acoustics.

Singing in the Shower

Perhaps the most familiar example of room resonance is that produced by someone singing in a tiled shower. The singer, who may be impressed by his vocal power, is not hearing his true singing voice; he is primarily exciting, or activating, the resonance, or natural, frequencies of a highly resonant chamber. The resonance frequencies for a shower stall, or any rectangular room, are determined by its dimensions and those of its occupant. For simplicity in calculating the resonance frequencies I shall assume that a glass door completely

THEATER AT EPIDAURUS is widely regarded as the most beautiful in Greece. Although the stage no longer exists, nearly every seat is intact. The regular spacing of the risers behind the seats creates unusual sound reflections as illustrated on the next page.

closes the entrance to the shower and ignore the presence of the occupant.

My shower is three feet square and eight feet high, and I have demonstrated to my own satisfaction that my presence in it does not appreciably alter the low-frequency resonances. One can regard the enclosure as a kind of organ pipe, eight feet long and closed at both ends. The fundamental tone, or lowest frequency of vibration, of such an organ pipe is a tone whose wavelength is twice the length of the pipe, or 16 feet. To find the frequency of a sound of this wavelength one divides 1,125 feet (the speed of sound in air at 68 degrees Fahrenheit) by 16 feet, which yields almost exactly 70 cycles per second. As every music student knows, such an eight-foot organ pipe also generates a whole series of harmonic overtones that have frequencies of two, three, four, five and so on times the fundamental frequency, corresponding to 140, 210, 280, 350 and so forth cycles per second.

These are not, however, the only resonance frequencies in a three-dimensional shower; there are also transverse modes of vibration with their appropriate resonances and harmonics. It turns out that there is a triply infinite series of resonance frequencies. They are determined by the dimensions of the shower, the velocity of sound and the appropriate assignment of integral numbers in groups of three, corresponding to the three dimensions of the shower. The first seven members of the triplet series are 0, 0, 1; 0, 0, 2; 1, 1, 0; 0, 2, 0; 2, 0, 0; 1, 1, 1 and 1, 1, 2. The three digits in each triplet are integers, two of which may be zero, and each of the three may increase (theoretically) to infinity.

The first four resonance frequencies for my shower, calculated by the appropriate formulas, are 70, 140, 187 and 264 cycles per second. These are only the first four of the triply infinite

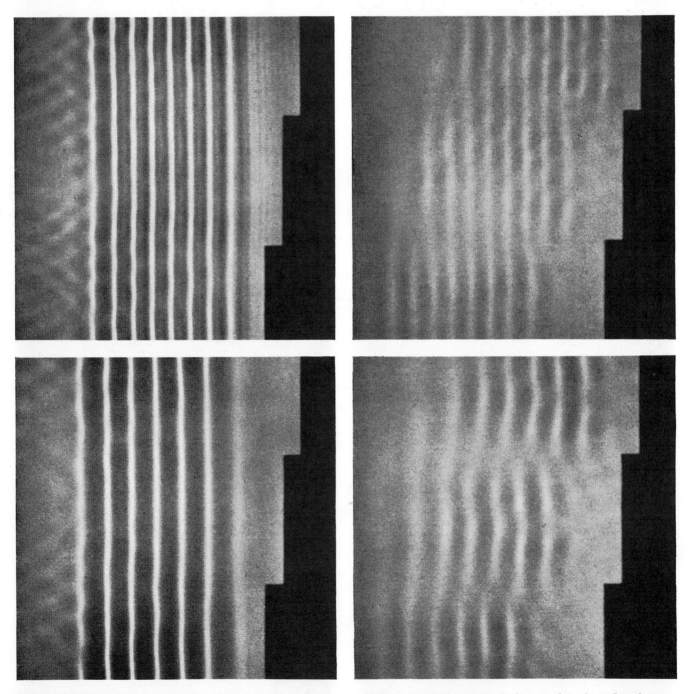

WAVE REFLECTION from regularly spaced risers, as in Greek open-air theaters, can be demonstrated with water waves in a ripple tank. When setback of risers is equal to integral multiples of a half-wavelength (*top pair of photographs*), the reflected waves are in phase. When the setback is equal to odd multiples of a quarter-wavelength (*bottom*), the reflected waves are out of phase.

series of resonance frequencies for this simple enclosure. At higher frequencies the separate modes of the resonant vibrations come closer and closer together and ultimately can no longer be resolved either by ear or by instrument. Therefore at sufficiently high frequencies this enclosure (or any rectangular room) has a frequency response that is essentially "flat," which means that it responds to all frequencies alike.

The prominence of resonance frequencies in a room is dependent on the reflective properties of its walls. Bathroom tile, for instance, reflects about 98 per cent of the sound energy that strikes it. Con-

sequently the resonance frequencies in a tiled shower are very prominent; moreover, the small dimensions of the shower give rise to resonances that have frequencies well within the audible range. In contrast, the prominent resonances in large rooms occur at frequencies that are usually below that range. Resonance frequencies can readily be suppressed by placing sound-absorptive materials on the wall surfaces of a resonant room. With nothing more than three large terry-cloth towels one can reduce the resonances in a tiled shower to the point where they are barely noticeable. This is done by placing one towel on the

floor and centering the other two on adjacent walls.

More than 30 years ago I investigated the resonances in a special experimental room eight feet square and 9.5 feet high. The room had concrete walls 10 inches thick and contained only one opening, which was sealed by a steel door. The first four resonance frequencies for this room, calculated for a temperature of 70 degrees F., were 59.2, 70.3, 92.9 and 99.7 cycles per second. These calculated values agreed with the experimentally determined frequencies quite precisely.

The top illustration on the next page shows how this room responds to sounds

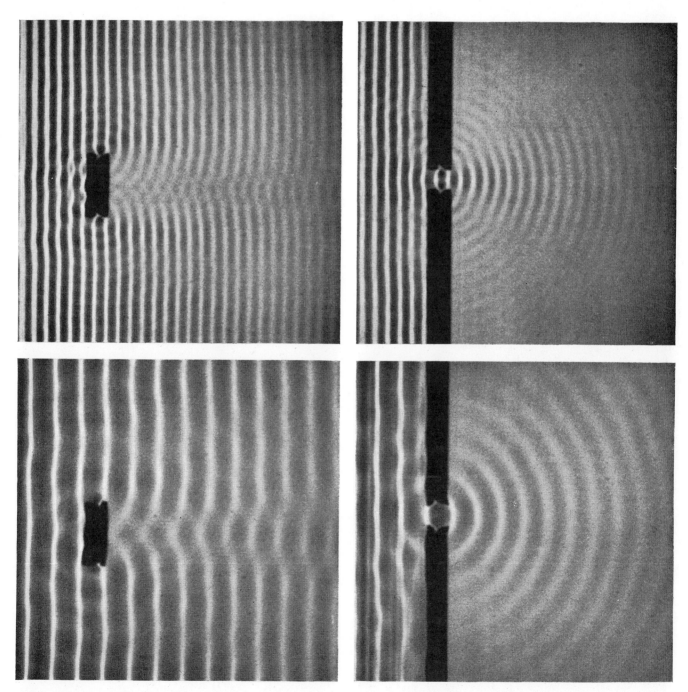

WAVE DIFFRACTION occurs when waves pass around an object or through an opening. When the wavelength is small compared with the size of the object or hole, the object tends to create a fairly sharp "shadow" (*top left*) and the waves tend to emerge from the hole in a "beam" (*top right*). As the wavelength is increased (*bottom*), the waves tend to spread more in both cases.

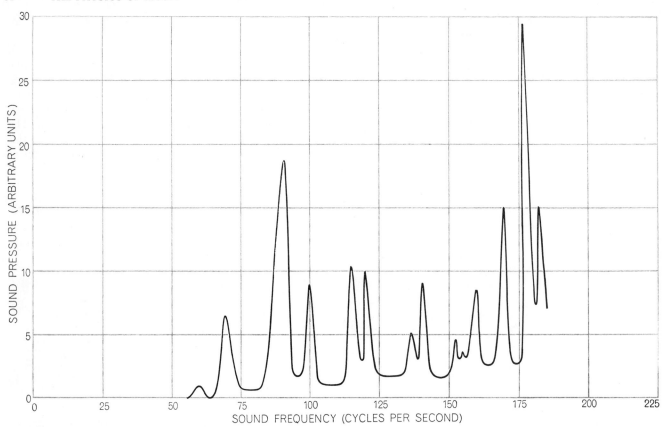

RESONANCE PEAKS were determined by the author for a massive concrete room eight feet square and 9.5 feet high. The curve shows how the room responds to sounds that have a frequency range between 50 and 185 cycles per second. The resonance peaks agree with those calculated. The peaks are proportional to the linear deflection of an oscillograph, therefore indicate sound pressure.

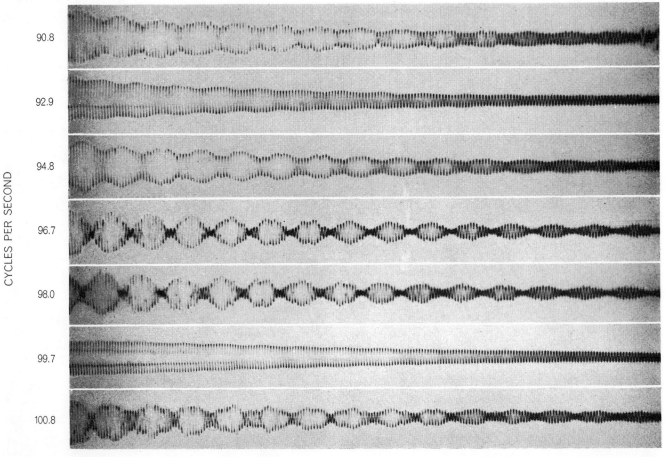

DECAY OF TONES in the room described above is shown in these oscillograms. Sounds of 92.9 and 99.7 cycles per second, which coincide with room resonance frequencies, decay smoothly. Tones of other frequencies do not. When the room is stimulated with a tone not a resonance frequency, the tone stimulates two or more resonance frequencies, which decay together and give rise to beats.

that have a frequency range between 50 and 185 cycles per second. It shows resonance peaks not only at the four frequencies listed above but also at nine higher frequencies, all of which agree with calculated values. The amplitudes of the peaks are proportional to the linear deflection of an oscillograph; hence they are more prominent than they would be if they were converted to decibels, which are based on a logarithmic scale. (On the decibel scale each interval of 10 units corresponds to a tenfold variation in sound energy. Thus a 100-decibel sound contains a million times more energy than a 40-decibel sound. To the ear, which hears a 10-decibel increase as an approximate doubling in strength, the 100-decibel sound is about 2^6, or 64, times louder than the 40-decibel sound.)

The bottom illustration on the opposite page shows how pure tones in the frequency range between 90 and 100 cycles per second die away in the same experimental room. In this range there are two prominent room resonances, at 92.9 and 99.7 cycles per second. Inspection of the seven oscillograms shows that sounds of these two frequencies decay smoothly, whereas sounds that do not coincide with resonance frequencies decay irregularly. Analysis of the five irregular decay oscillograms reveals that all are made up of two or more of the resonance frequencies of the room. For example, the fourth oscillogram consists of the decay of the two resonance frequencies of 92.9 and 99.7 cycles per second. It is apparent that when the room is stimulated with a tone that has a frequency about halfway between these two resonance frequencies, both frequencies are stimulated and decay together. In the process they give rise to "beats" with a frequency of 6.8 beats per second, which is precisely the frequency difference between the two resonance frequencies $(99.7 - 92.9 = 6.8)$.

These tone-decay oscillograms demonstrate convincingly that room reverberation is made up of the free decay of the room's resonance frequencies, or natural modes of vibration. Such a phenomenon is readily understood in terms of wave acoustics but is wholly unpredictable by ray methods. For lack of understanding of wave acoustics many architects believe that optimum reverberatory properties of a room can be obtained by treating one surface of the room, usually the ceiling, with an absorptive acoustical material ("acoustic tile"). Although such a treatment is usually beneficial in large public rooms such as restaurants and offices, where the prime objective is to reduce the general noise level, it is often

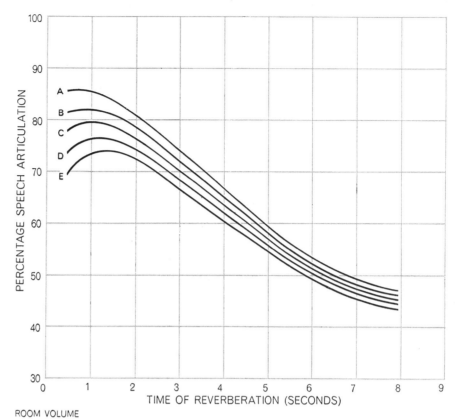

ROOM VOLUME
(CUBIC FEET)

A 25,000
B 100,000
C 400,000
D 800,000
E 1,600 000

AUDIBILITY OF SPEECH in a quiet room depends primarily on reverberation time and size of room. Audibility is expressed as Percentage Speech Articulation (PA), the per cent of unamplified speech sounds that a panel of listeners identifies correctly. Curves show the difficulty of achieving a PA above 75 per cent, the minimum acceptable value.

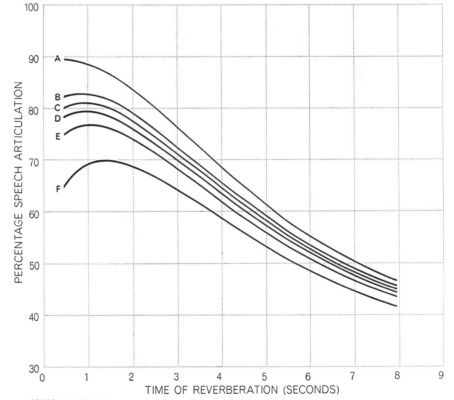

A SPEECH AMPLIFIED
B LOUDEST SPEAKER
C MODERATELY LOUD
D AVERAGE
E MODERATELY WEAK
F WEAKEST SPEAKER

AUDIBILITY OF SPEAKERS depends on their vocal strength and reverberation time. The curves show the Percentage Speech Articulation for amplified and unamplified speech in an auditorium of 400,000 cubic feet. Both sets of curves are by the author.

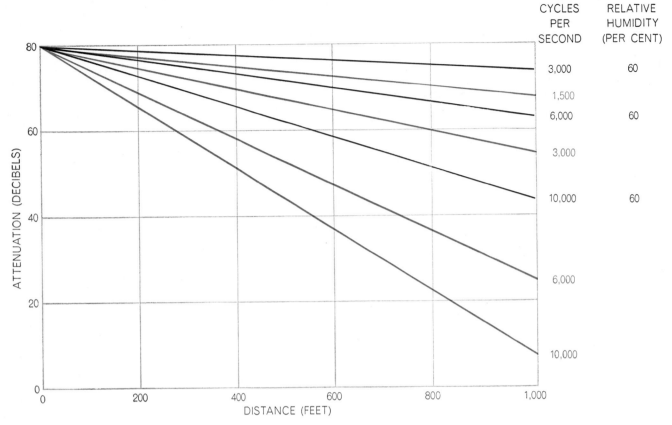

SOUND ABSORPTION IN AIR varies with frequency, humidity and temperature. Black curves show sound attenuation at typical concert-hall humidity, 60 per cent, and at 70 degrees Fahrenheit. Colored curves show attenuation at humidities for which sound absorption is highest, at 70 degrees F. Each 20-decibel drop represents a decrease by a factor of 100 in sound energy. Thus music heard outdoors on a very dry, warm night will sound deficient in high frequencies, particularly at distances over a few hundred feet.

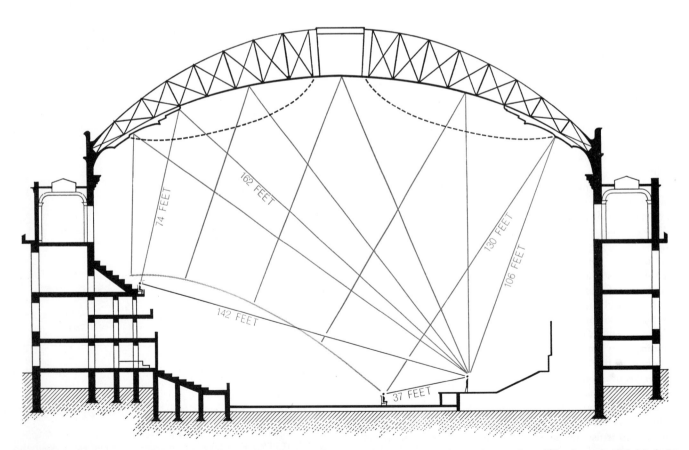

ROYAL ALBERT HALL, opened in 1871, was originally plagued by echoes reflected from the great dome. The colored lines show reflections of equal travel time. A listener in the front of the audi-torium would hear an echo nearly a fifth of a second behind the direct sound. Echoes and reverberation were much reduced by installation of a velarium, or heavy fabric awning (broken lines).

harmful in music rooms, and it is almost never a sufficient measure for obtaining good acoustical performance.

Interference and Diffraction

We have seen in the discussion of open-air theaters how regularly spaced reflecting surfaces can lead to deleterious reinforcement and cancellation of harmonic series of sound frequencies. Similar but usually less serious interference effects also take place indoors when sound waves encounter the boundaries of a room. The interference between the direct and reflected waves is aggravated if the room has prominent modes of resonant vibration. To minimize such difficulties the architect must avoid large, smooth reflecting surfaces and either judiciously introduce irregularities in the boundary contours of the room (for instance window frames, pilasters and niches for art objects) or install randomly placed panels of sound-absorptive materials on the large reflecting surfaces. The purpose is to attain a high degree of diffusion of the reflected sound so that everywhere in the room there will be a multitude of reflected sound waves coming from all directions and meeting in random phases.

Even more subtle than the effects produced by interference are those produced by diffraction of sound waves. Everyone has noticed how sound waves bend around corners. An automobile concealed by a building can be heard even when it cannot be seen. To shut out noise from a room one must close a door or window completely. The closing of the last half-inch often excludes more noise than was excluded by the entire closure up to that point.

Sound waves, like light waves, bend or spread around an obstacle when the dimensions of the obstacle are comparable to the wavelength. The waves pass around the obstacle and unite in various combinations of phases, yielding the familiar diffraction patterns. The same kinds of patterns arise when waves pass through a small hole. The sound energy emerging through such a hole is often much more than would be indicated by multiplying the incident sound energy per unit area (a wavelength or more in front of the hole) by the area of the hole.

Because audible sound waves cover such a broad spectrum in size, the diffraction patterns produced by a given obstacle (or hole) can be exceedingly complex. Moreover, a hard (reflective) obstacle that is, say, two feet across will almost totally reflect high-frequency

sounds, which have wavelengths measured in inches, and, if it is adjacent to an opening of comparable size, the obstacle and opening (acting together) will be almost totally transparent to low-frequency sounds, which have wavelengths measured in tens of feet.

One of the architect's problems is that in trying to correct one kind of acoustical defect he may introduce others. This is particularly true of the use of suspended ceiling panels, which are useful for overcoming such acoustical defects as echoes and long-delayed reflections. Suspended panels have been used to good advantage in several important concert halls and auditoriums, among them the Stockholm Concert Hall and the Tanglewood Music Shed in Lenox, Mass. There have also been less successful examples. The difficulty is that sounds of short wavelength are almost completely reflected by the panels, whereas sounds of long wavelength pass almost completely around the panels, to be reflected later by the ceiling above. For most arrays of panels the transmission, reflection, diffraction and scattering effects are greatly dependent on the wavelength of the sound. Precise calculation of these effects by the methods of wave acoustics is possible

only for regularly spaced circular or rectangular panels, and even such simple arrays require formidable calculations. Where such calculations cannot be made it is imperative that the acoustical effects be studied in three-dimensional models using sound waves whose wavelengths are in the same ratio to the model as actual sound waves are to the full-sized room. Thus if the model is one twenty-fourth actual size, the experimental sound waves must be reduced in length correspondingly. In such a model one would use sound waves with a frequency of 24,000 cycles per second to simulate the effect of 1,000-cycle-per-second waves in a full-sized room.

The behavior of these high-frequency waves, which are inaudible, can be studied in a number of ways. One method is to photograph the waves produced by an electric spark [*see illustration on pages 78 and 79*]. Another method, recently developed in Germany, is to record in the model the high-frequency sounds on magnetic tape and play them back at reduced speed so that they can be heard as they would sound in a full-sized version of the room.

Manfred R. Schroeder of the Bell Telephone Laboratories has developed a promising method of using an elec-

ANECHOIC CHAMBER is employed by Manfred R. Schroeder of Bell Telephone Laboratories to evaluate stereophonic playback of a sound recording that has been modified by a computer to simulate the acoustics of a newly designed music room or concert hall.

tronic computer to simulate the acoustics of planned music rooms. He has devised computer programs that will modify a tape recording of music so that the music sounds as if it were being played in the room under study. For appraisal the tape is played back through multiple loudspeakers in a special anechoic, or echoless, chamber [*see illustration on preceding page*]. Whatever method one uses, it is essential that the performance of suspended panels and all other acoustical innovations be fully explored in advance of installation.

When the architect faces the job of designing a new concert hall or other music room, he must pay primary attention to three things: the shape of the total enclosure, the design of the stage and music shell and the reverberation time. A felicitous shape is a requirement of the highest priority. Unfortunately many architects believe that faulty shapes can be corrected by covering the offending surfaces with highly absorptive materials and by adjusting the reverberation time. Thus deluded, they adopt a fash-

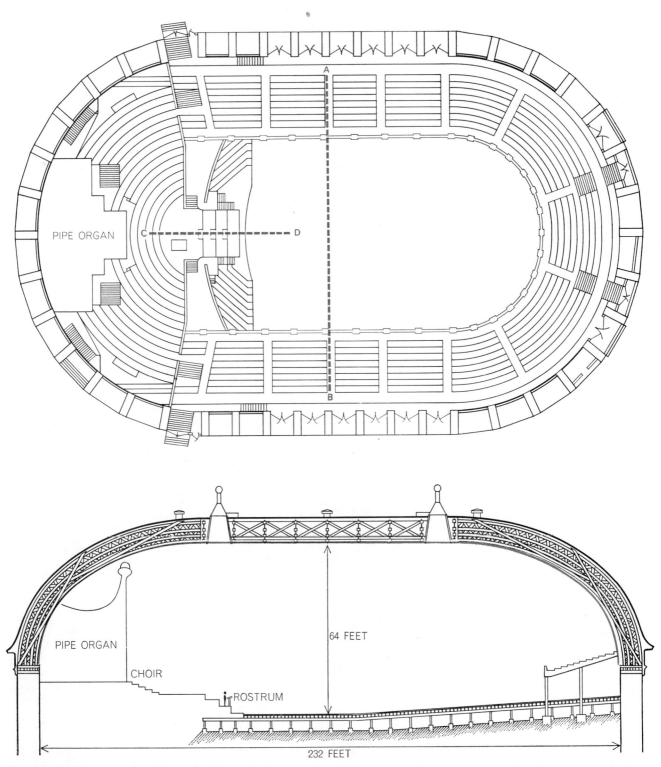

MORMON TABERNACLE, built in Salt Lake City about 100 years ago, has generally fine acoustics, particularly when the audience is about 2,500, a number that provides optimum reverberation time. In the plan view *A* and *B* show the locations of the pistol and micro- phone used to make the upper decay curve on the opposite page. *C* and *D* show pistol and microphone locations used in making the lower decay curve. The colored line in the sectional view (*bottom*) indicates where the builders used plaster containing cattle hair.

ionable construction method, such as the concrete shell, and produce a building that is an acoustical perversion. A bad shape is a permanent liability.

It is not difficult to arrive at acoustically satisfactory shapes using the methods of ray and wave acoustics. Experience and ratings by competent listeners indicate that for generally rectangular rooms that have volumes between 15,000 and 500,000 cubic feet a favorable ratio of length to width is about four to three, and a good ceiling height is about .6 times the cube root of the volume. Therefore for a chamber-music room seating about 200 people the favorable dimensions would be about 52 by 40 by 20 feet. It is usually desirable to make the side walls diverge slightly and to incline the floor so that it is not parallel to the ceiling. The reason is that parallel surfaces have a tendency to produce flutter echoes. If opposing surfaces must be kept parallel for some reason, flutter echoes can be suppressed by the use of diffusive or absorptive panels. Whenever possible, particularly in large music rooms, the design should be subjected to a thorough wave-acoustics analysis to ensure that the room will not be impaired by ill effects of resonance, interference and diffraction.

Reflection and Reverberation

An important consideration in music rooms is the time delays in the successive reflections reaching listeners seated in various parts of the room. For rooms that have volumes between 150,000 and 400,000 cubic feet the first reflections should be delayed not more than about 30 to 35 milliseconds beyond the arrival of the direct sound, and these first reflections should be followed by a succession of reflections, coming from all directions, that will "envelop" the listeners with a relatively smooth but slightly undulating reverberation (much like a vibrato) of the optimum duration and frequency characteristic. For larger rooms the first reflections should be delayed not more than about 45 milliseconds; the reflections should be diffuse and should come in good proportions from the side walls, the rear wall and the ceiling.

The acoustical design of the stage enclosure must meet two general requirements. First, the reverberation characteristic of the stage space, with its normal hangings and equipment, should not differ appreciably from that of the audience space. Second, there must be a properly

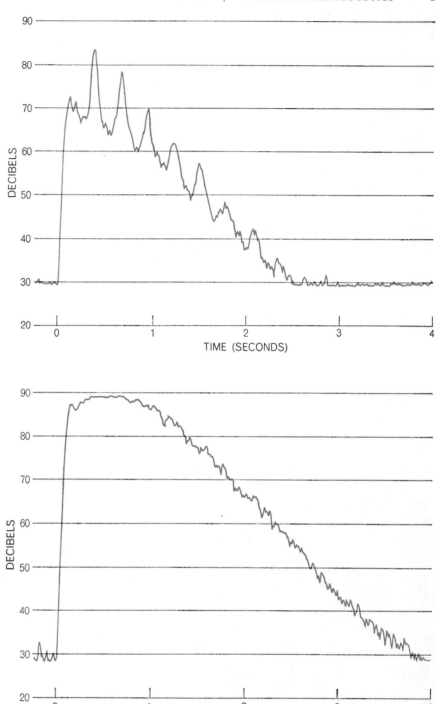

ECHOES IN MORMON TABERNACLE, recorded by Harvey Fletcher, William L. Woolf and the author, were produced by pistol shots at two different locations. The top curve shows the flutter echo when the pistol is fired in one balcony and recorded in an opposite balcony (*see upper illustration on opposite page*). A smoothly decaying reverberation (*bottom*) results when a pistol is fired on the rostrum and recorded part-way back in the Tabernacle.

designed music shell that will enable all the members of an orchestra to hear each other clearly and distinctly, blend and unify the sound of the entire ensemble and reflect a large portion of this enhanced sound to the audience.

The subject of reverberation time, the third major aspect of music-room design, has been given extensive study by acousticians, beginning more than 60 years ago with the pioneer work of Wallace C. Sabine of Harvard University. As we have seen, reverberation in a room is the persistence of the natural modes of vibration, or resonance frequencies, after the source of sound in the room has been stopped. For acoustical purposes reverberation is defined as the time required for the persistent sound to decay, or diminish, by 60 decibels, which is an energy factor of a million.

Experience has shown that the optimum reverberation times for sound of 1,000 cycles per second are as follows: .5 second for small practice rooms (volume about 500 cubic feet); .8 to one second for rehearsal rooms (volumes up to 15,000 cubic feet); 1.1 to 1.4 seconds for chamber-music rooms (volumes between 35,000 and 75,000 cubic feet); 1.7 to two seconds for large concert halls (volumes between 350,000 and 700,000 cubic feet), and about two to 2.2 seconds in very large halls used for organ music and choral works. In Europe habit and experience give preference to reverberation times about 10 per cent longer than these.

Acousticians have also given much thought to the problem of how the reverberation time should vary with frequency. Should it be the same for all frequencies? Should it be based on the frequency distribution of sound energy in music, so that on the average all components will die away to inaudibility in the same length of time? Or should it be such that the rate of growth or de-

cay of loudness level will be the same for all frequency components? Fortunately the last two criteria lead to about the same reverberation characteristics for frequencies below 1,000 cycles per second: a gently rising one in which the reverberation time is about 50 per cent longer at 62 cycles per second than it is at 1,000 cycles. Thus if the optimum reverberation time for 1,000 cycles is two seconds, that for 62 cycles is three seconds.

In order to obtain the desired reverberation times the designer of a music room has available a wide selection of building and decorative materials: stone, brick, wood, plaster, natural and synthetic fabrics and a great variety of acoustic tiles and composition panels. Their sound-absorption coefficients have been determined by careful measurements in reverberation chambers. By constructing the interior surfaces of a music room with judicious combinations of materials it is possible to obtain optimum reverberation times for all important frequencies throughout the audible range. Two recent examples of music rooms designed to meet these criteria

are Hertz Hall on the Berkeley campus of the University of California and the Seattle Opera House.

Two Older Music Halls

I shall mention briefly two famous structures that were designed before acoustical knowledge was applied to architectural design: the Royal Albert Hall in London and the Mormon Tabernacle in Salt Lake City. The former, opened in 1871, is interesting because it exhibits nearly all the acoustical defects that should be avoided in the design of concert halls. The bottom illustration on page 86 shows several of the long-delayed echoes that result from the high, domed ceiling. Sound reflected from the ceiling can be delayed nearly a fifth of a second behind the direct sound and, because of the focusing effect of the ceiling, it can be nearly as loud. These extremely disturbing echoes have been greatly reduced by suspending a convex velarium, or canopy, below the ceiling, which also helps to reduce another defect of the hall: excessive reverberation.

In contrast, the Mormon Tabernacle,

HERTZ HALL of the University of California at Berkeley is an example of an auditorium that meets high acoustical standards. The forward stage can be elevated to take an entire symphony orchestra. Absorptive panels on side walls provide good diffusion and proper reverberation. The architect was Gardner A. Dailey and Associates. The acoustical consultants were Delsasso and the author.

completed in 1867, is famous for its good acoustics. Its virtues are somewhat surprising considering that its floor plan is approximately elliptical and the high, domed ceiling is elliptical in its transverse section [*see illustration on page 88*]. Fortunately the convergent reflections from the ceiling are disturbing at only a few locations. For the most part the concave surfaces around and above the organ and choir area sustain and blend the instrumental and vocal sounds and project them most effectively throughout the auditorium.

A recent study I have made in association with Harvey Fletcher and William L. Woolf has shown that the elliptical ceiling gives rise to a very prominent flutter echo when a sharp sound is made along the upper side balconies. The echo, set off by a pistol shot, is shown in the illustration on page 89. Fortunately this flutter echo is not activated appreciably by sounds originating in the organ and choir area.

Except for the floors, which are wood, the interior surfaces of the Tabernacle are mostly lime plaster on wood lath. Today such construction would produce an excessively long reverberation time, but at the time the Tabernacle was built it was the custom to mix large amounts of cattle hair with the plaster. This made the plaster considerably more absorptive than it would have been otherwise. Even so, when there is no audience, the reverberation time is about four seconds at 1,000 cycles per second. The optimum time for the Tabernacle, when music is being played, would be about 2.2 seconds.

It happens that the optimum reverberation time is obtained almost exactly when there is an audience of some 2,500. With an audience of about 6,500 the reverberation time drops to slightly below 1.5 seconds at 1,000 cycles per second. The Tabernacle management and radio station KSL frequently get letters of praise for the quality of music broadcast when the Tabernacle contains an audience of approximately 2,500. Complaints about the acoustics are often received, however, when the audience numbers 6,000 or more.

Compared with designing a music room, the job of designing a room used primarily for speech (for example lecture rooms and legislative rooms) is relatively simple. Like a music room, a speech room should be free from external sources of noise, from resonances and from echoes and sound-focusing effects. The reverberation time should be about a second (slightly less for small rooms and slightly more for large rooms). Finally, speech must be clearly heard throughout the room. This means that unless the room is small an amplifying system must be provided.

The general need for amplification can be demonstrated by using speech-articulation tests of the type developed by telephone engineers to determine the intelligibility of transmitted speech. The results are expressed in terms of Percentage Speech Articulation (PA). The PA is determined by having typical speakers call out speech sounds that occur in English words and having a panel of listeners record what they think they hear. If they hear correctly 75 per cent of the speech sounds, the PA is 75 per cent, which is the minimum value acceptable for satisfactory hearing.

The illustration at the top of page 85 shows in a family of curves how the

SEATTLE OPERA HOUSE has many acoustical features. Rear walls at all levels are inclined forward to prevent echoes and reflect sound beneficially. Wood panels on side walls can be opened to expose absorptive chambers that change reverberation time. The architects were B. Marcus Priteca and James J. Chiarelli. The acoustical consultants were Paul S. Veneklasen and the author.

hearing of the unamplified speech of the average speaker depends on the time of reverberation and room size. It will be seen that except for the smallest rooms with ideal reverberation times the PA values are below 75 per cent. The conclusion is that amplification is almost always necessary.

The illustration at the bottom of page 85 shows the PA for 14 speakers of widely different vocal strengths in an auditorium of 400,000 cubic feet. The curves indicate clearly why some speakers are heard much better than others and why some are not heard at all, and they demonstrate once again why it is generally advantageous to amplify speech.

The Acoustics of Homes

Perhaps the biggest failure of U.S. architects (and acousticians) is in not doing something constructive about the acoustical environment of the home, particularly the apartment dwellings that have been built in the past 20 years. An easily attained objective would be to shorten the reverberation time of small rooms to about .5 second and that of large living rooms used for music as well as speech to no more than one second. Typical values in U.S. homes with con-

ventional plaster walls and ceiling, and with scant carpeting, are from 50 to 100 per cent higher. The more urgent and difficult problem is to screen out unwanted noise, whether it is of external or internal origin. (The flushing of a toilet makes a racket that often carries through every room of the house.)

U.S. cities should adopt as construction standards the many commendable and necessary acoustical features that are to be found in virtually all new apartments in both eastern and western Europe. Apartments are carefully planned so that rooms in which the loudest noises are likely to originate are the farthest from those in which the most quiet is desired. There is, for example, maximum separation between the bedroom of one apartment and the living room of the adjacent one. There is a heavy wall between adjacent bathrooms; the entrance hall is used as a sound lock between the living room and the bedroom; entrance doors are of solid-panel construction, well fitted in their frames so that threshold cracks are eliminated. Floors above ground level are usually floating concrete slabs, so that impact sounds as well as air-borne ones are thoroughly insulated.

The effective control of noise in these modern European buildings is no acci-

dent. It is required by the high standards of building codes. Sweden, for instance, requires that there be enough sound insulation between rooms in residential buildings to reduce air-borne sounds by 48 decibels. Even higher standards are required in hospitals and certain other buildings. In the U.S., in contrast, sound insulation is completely ignored by practically all building codes. (It is a hopeful sign that a new code being considered for New York City may include noise-level specifications for the first time.)

The provision of quiet buildings, particularly those in which people live, learn and recover from illness, is an essential objective of good community planning. A nation that prides itself on its high technology and high standard of living should be willing to pay the additional 5 or 10 per cent that would be entailed in creating buildings with satisfactory acoustics.

I am sure it can be demonstrated that good acoustics is good business, but far more important, the providing of quiet buildings is indispensable for good health and the growth of culture. Home should be our refuge from a noisy world, where taut nerves may find rest from and refreshment for the strains of high-pressure living.

HOLLYWOOD BOWL, noted for its acoustics, was opened in 1922. The shell over the stage was an architectural and acoustical innovation, since much copied. The architect was Lloyd Wright, son of Frank Lloyd Wright. The author was acoustical consultant.

BIBLIOGRAPHIES

1. Physics and Music

THE PHYSICS OF MUSIC. Alexander Wood. Methuen, 1947.

SOUND WAVES: THEIR SHAPE AND SPEED. D. C. Miller. Macmillan, 1937.

THE PSYCHOLOGY OF MUSIC. C. E. Seashore. McGraw-Hill, 1938.

THE ACOUSTICS OF MUSIC. W. T. Bartholomew. Prentice-Hall, 1942.

VISIBLE SPEECH. R. K. Potter and others. Van Nostrand, 1948.

2. The Acoustics of the Singing Voice

SINGING: THE MECHANISM AND THE TECHNIC. William Vannard. Carl Fischer, Inc., 1967.

TOWARDS AN INTEGRATED PHYSIOLOGIC-ACOUSTIC THEORY OF VOCAL REGISTERS. John Large in *NATS Bulletin*, Vol. 28, No. 3, pages 18–25, 30–36; February/March, 1972.

ARTICULATORY INTERPRETATION OF THE "SINGING FORMANT." Johan Sundberg in *The Journal of the Acoustical Society of America*, Vol. 55, No. 4, pages 838–844; April, 1974.

FUNDAMENTALS OF MUSICAL ACOUSTICS. Arthur H. Benade. Oxford University Press, 1976.

3. The Physics of the Piano

NORMAL VIBRATION FREQUENCIES OF A STIFF PIANO STRING. Harvey Fletcher in *The Journal of the Acoustical Society of America*, Vol. 36, No. 1, pages 203–209; January, 1964.

QUALITY OF PIANO TONES. Harvey Fletcher, E. Donnell Blackham and Richard Stratton in *The Journal of the Acoustical Society of America*, Vol. 34, No. 6, pages 749–761; June, 1962.

THE THEORY OF SOUND: VOL. I. John William Strutt, Baron Rayleigh. Macmillan and Co., Limited, 1929.

VIBRATION AND SOUND. Philip M. Morse. McGraw-Hill Book Company, Inc., 1948.

4. The Physics of Wood Winds

THE ACOUSTICS OF ORCHESTRAL INSTRUMENTS AND OF THE ORGAN. E. G. Richardson. Arnold & Co., 1929.

HORNS, STRINGS AND HARMONY. Arthur H. Benade. Doubleday & Company, Inc., 1960.

ON WOODWIND INSTRUMENT BORES. A. H. Benade in *The Journal of the Acoustical Society of America*, Vol. 31, No. 2, pages 137–146; February, 1959.

THE PHYSICAL BASIS OF MUSIC. Alexander Wood. Cambridge University Press, 1913.

WOODWIND INSTRUMENTS AND THEIR HISTORY. Anthony Baines. W. W. Norton & Company, 1957.

5. The Physics of Brasses

HORNS, STRINGS AND HARMONY. Arthur H. Benade. Doubleday & Company, Inc., 1960.

THE TRUMPET AND TROMBONE. Philip Bate. W. W. Norton & Company, Inc., 1966.

COMPLETE SOLUTIONS OF THE "WEBSTER" HORN EQUATION. Edward Eisner in *The Journal of the Acoustical Society of America*, Vol. 41, No. 4, Part 2, pages 1126–1146; April, 1967.

ON PLANE AND SPHERICAL WAVES IN HORNS OF NONUNIFORM FLARE. E. V. Jansson and A. H. Benade. Technical Report, Speech Transmission Laboratory, Royal Institute of Technology, Stockholm, March 15, 1973.

6. The Physics of Violins

THE MECHANICAL ACTION OF INSTRUMENTS OF THE VIOLIN FAMILY. F. A. Saunders in *The Journal of the Acoustical Society of America*, Vol. 17, No. 3, pages 169–186; January, 1946.

THE MECHANICAL ACTION OF VIOLINS. F. A. Saunders in *The Journal of the Acoustical Society of America,* Vol. 9, No. 2, pages 81–98; October, 1937.

MISURA DELL'ATTRITO INTERNO E DELLE COSTANTI ELASTICHE DEL LEGNO. I. Barducci and G. Pasqualini in *Nuovo Cimento,* Vol. 5, No. 5, pages 416–446; October 1, 1948.

REGARDING THE SOUND QUALITY OF VIOLINS AND A SCIENTIFIC BASIS FOR VIOLIN CONSTRUCTION. H. Meinel in *The Journal of the Acoustical Society of America,* Vol. 29, No. 7, pages 817–822; July, 1957.

SUBHARMONICS AND PLATE TAP TONES IN VIOLIN ACOUSTICS. Carleen M. Hutchins, Alvin S. Hopping and Frederick A. Saunders in *The Journal of the Acoustical Society of America,* Vol. 32, No. 11, pages 1443–1449; November, 1960.

7. The Physics of the Bowed String

ON THE MECHANICAL THEORY OF VIBRATIONS OF BOWED STRINGS AND OF MUSICAL INSTRUMENTS OF THE VIOLIN FAMILY. C. V. Raman. Indian Association for the Cultivation of Science, 1918.

THE MECHANICAL ACTION OF VIOLINS. F. A. Saunders in *The Journal of the Acoustical Society of America,* Vol. 9, No. 2, pages 81–98; October, 1937.

ON THE SENSATIONS OF TONE AS A PHYSIOLOGICAL BASIS FOR THE THEORY OF MUSIC. Hermann L. F. Helmholtz. Dover Publications, Inc., 1954.

THE BOWED STRING AND THE PLAYER. John C. Schelleng in *The Journal of the Acoustical Society of America,* Vol. 53, No. 1, pages 26–41; January, 1973.

8. Architectural Acoustics

ACOUSTICAL DESIGNING IN ARCHITECTURE. V. O. Knudsen and Cyril M. Harris. John Wiley & Sons, Inc., 1950.

ACOUSTICS, NOISE AND BUILDINGS. P. H. Parkin and H. R. Humphreys. Faber and Faber Ltd, 1958.

ARCHITECTURAL ACOUSTICS. Vern O. Knudsen. John Wiley & Sons, Inc., 1932.

MUSIC, ACOUSTICS AND ARCHITECTURE. Leo L. Beranek. John Wiley & Sons, Inc., 1962.

INDEX